Drones

This book is part of the Peter Lang Media and Communication list.
Every volume is peer reviewed and meets
the highest quality standards for content and production.

PETER LANG
New York • Bern • Frankfurt • Berlin
Brussels • Vienna • Oxford • Warsaw

Kevin Howley

Drones

Media Discourse and the Public Imagination

PETER LANG
New York • Bern • Frankfurt • Berlin
Brussels • Vienna • Oxford • Warsaw

Library of Congress Cataloging-in-Publication Data

Names: Howley, Kevin, author.
Title: Drones: media discourse and the public imagination / Kevin Howley.
Description: New York: Peter Lang, 2018.
Includes bibliographical references and index.
Identifiers: LCCN 2017032398 | ISBN 978-1-4331-2640-6 (pbk.: alk. paper)
ISBN 978-1-4331-4741-8 (hardback: alk. paper)
ISBN 978-1-4331-4742-5 (ebook pdf)
ISBN 978-1-4331-4743-2 (epub) | ISBN 978-1-4331-4744-9 (mobi)
Subjects: LCSH: Drone aircraft. | Discourse analysis—Social aspects. |
Mass media—Social aspects.
Classification: LCC UG1242.D7 H69 | DDC 358.4/183—dc23
LC record available at https://lccn.loc.gov/2017032398
DOI 10.3726/b11986

Bibliographic information published by **Die Deutsche Nationalbibliothek.**
Die Deutsche Nationalbibliothek lists this publication in the "Deutsche Nationalbibliografie"; detailed bibliographic data are available on the Internet at http://dnb.d-nb.de/.

An earlier version of Chapter 11 appears in Howley, K. (2016). "I Have a Drone": Internet memes and the politics of culture. *Interactions: Studies in Communication & Culture*, *7*(2), 155–175.

Chapter 12 first appeared in Howley, K. (2017). Drone warfare: Twenty-first century empire and communications. *MediaTropes*, 6(2), 45–62.

The paper in this book meets the guidelines for permanence and durability of the Committee on Production Guidelines for Book Longevity of the Council of Library Resources.

29 Broadway, 18th floor, New York, NY 10006
www.peterlang.com

Printed in the United States of America

To the innocents and those who bear witness.

CONTENTS

FIGURES

PREFACE

Books about drones are nearly as commonplace—if not quite so infamous—as these flying machines themselves. In recent years, well-known authors and critics have produced timely analyses of the drone wars from activist, historical, and international relations perspectives. For instance, Medea Benjamin, co-founder of the progressive grassroots organization, Code Pink, published *Drone Warfare: Killing By Remote Control.* Political reporters Nick Turse and Tom Engelhardt co-authored *Terminator Planet: The First History of Drone Warfare 2001–2050*. Lloyd C. Gardner's *Killing Machine: The American Presidency in the Age of Drone Warfare* examines the implications of weaponized drones for presidential power in the twenty-first century. And America's most distinguished public intellectual, Noam Chomsky, is co-author, with Andre Vltchek, of *On Western Terrorism: From Hiroshima to Drone War*. Other titles deal with the legal, strategic, and policy implications of drone warfare. Three notable examples: Brian Glyn Williams's *Predators: The CIA's Drone War on al Qaeda*; Akbar Ahmed's *The Thistle and the Drone: How America's War on Terror Became a Global War on Tribal Islam*; and Marjorie Cohn's *Drones and Targeted Killing: Legal, Moral & Geopolitical Issues*.

At the same time, researchers from a variety of disciplines—engineering, computer science, international relations, cultural studies, and political science—have published scholarly analyses of drones in peer-reviewed journals. Conversely, drones, robotics, and automated killing machines feature prominently in recent popular fiction, from Mike Maden's *Drone (A Troy Pearce Novel)* to former counterterrorism czar Richard Clarke's political thriller *Sting of the Drone*. Then there's Don Nardo's *Drones (Military Experience: In the Air)*, which offers a historical account of weaponized drones for young readers. Drones have even made an appearance in the self-help genre; see, for example, J. J. Best's *UAV Pilot: How to Be Ready for the Coming Drone Pilot Job Boom*. And just in time for the gift-giving season, Craig S. Issod authored *Getting Started with Hobby Quadcopters and Drones*.

Despite growing academic and popular interest, there are no book-length treatments of drones from media and cultural studies perspectives. Nevertheless, as I hope to demonstrate, operating from such a perspective reveals a great deal about the relationship between technology and culture. Specifically, my aim is to offer a comprehensive understanding of the role news stories, advertising, activist media, and popular culture play in constructing public knowledge of this emerging technology. In doing so, this book highlights how media discourse—talk and text, sound and image—shapes and is shaped by technological innovation and artifacts, like drones.

Research support, in the form of a sabbatical leave from DePauw University during the 2014–2015 academic year, was instrumental in getting this project off the ground. Funding through DePauw's interdisciplinary Conflict Studies program and a subsequent Faculty Fellowship award provided additional time and resources to bring the book to completion. Thanks to my friends and colleagues at DePauw, most notably, Istvan Csicsery-Ronay, Jeff Kenney, Tiffany Hebb, Jin Kim, Geoff Klinger, and especially, Jeff McCall, for their assistance and support as this project moved forward. Further afield, my gratitude goes out to Arthur Holland Michel and the faculty and students at the Center for the Study for the Drone at Bard College for their help during the early stages of my research. Likewise, I'm grateful to Øyvind Vågnes, at the University of Bergen, for his feedback and encouragement during my travels to the 2015 meeting of the American Studies Association of Norway. Thanks also to Mitch Reyes at Lewis & Clark College for organizing two highly productive writing retreats that proved instrumental in bringing this project to completion. Finally, I'd like to thank Kathryn Harrison, Michael Doub and their colleagues at Peter Lang for their professionalism and expertise.

INTRODUCTION

Don't Call Them Drones

Ultimately, the meaning of a tool is inseparable from the stories that surround it.
—David Nye, *Technology Matters: Questions to Live With*

A story in *Popular Mechanics* headlined "Drone Is Not a Four-Letter Word" describes a brief encounter between technology reporter, Joe Pappalardo, and a New York City police officer. Fresh from an appearance on the Fox Business Channel, where he discussed civilian and commercial uses of these suddenly ubiquitous flying machines, Pappalardo attempted to reassure the cop, a veteran of the Afghan War, that he has nothing to fear from unmanned aerial vehicles (UAVs), commonly known as drones. Unconvinced, the officer expressed his discomfort: "If I saw one of these while I was out on my boat, I'd just, you know, shoot it down" (Pappalardo, 2013). The policeman's unease, Pappalardo observes, is not unique. Sampling social media responses to Amazon's highly publicized plans for delivering packages by quadcopter, Pappalardo describes public skepticism toward the online retailer's lofty scheme, and more broadly, the source of profound cultural anxieties articulated by the rise of the drone. The journalist offers several reasons for this skeptical, frequently apprehensive, popular reaction to drones—concern over public safety and fear of automation, among others—before settling on a somewhat unexpected explanation: language. "I think there's a linguistic problem here, too, stemming from the use of the word 'drone.' The law enforcement and military connotations of the word runs deep," Pappalardo concludes. It is a plausible interpretation. After all, in the age of drone warfare, with its attendant Orwellian language –"imminent threat," "enemy non-combatant," "kinetic military action," "disposition matrix" and the like—words certainly do matter.

But this isn't the whole story. Evidently, "drone" is a four-letter word for the very same military personnel who routinely use these aircraft for reconnaissance, surveillance, and targeted killing. During a 2012 interview with NPR, outgoing US Air Force General Norton Schwartz dressed down reporter Tom Bowman for uttering the phrase "drone." "Drones mischaracterize what these things are," Schwartz declared. "They're not dumb. Nor are they unmanned, actually. They're remotely piloted aircraft" (Bowman, 2012). US Army General Martin Dempsey is equally perturbed by what journalist Aram Roston adroitly refers to as "the D word." Returning to Washington from a 2014 defense meeting at NATO headquarters in Brussels, Dempsey told reporters, "You will never hear me use the word 'drone,' and you'll never hear me use the term 'unmanned aerial systems,' because they are not. They are remotely piloted aircraft" (Garamone, 2014). With its emphasis on highly trained pilots in control of sophisticated, precision aircraft, the military's designation, better known by the acronym RPA, avoids the gendered nomination, unmanned aerial systems (UAS), preferred by entrepreneurs, trade organizations, and industry groups.

For instance, Ray Mann, a Welsh engineer and owner of "the world's first civilian airport for drones," was unequivocal when he told the *Sunday Times*, "Don't call them drones. I don't like the word. It makes people fearful. It has a sinister sound to it" (Gillespie, 2013). Likewise, Lucien Miller, CEO of Innov-8tive Designs in Vista, California, is keenly aware of the perception problem entrepreneurs face in light of "Predators and their Hellfire missiles bombing daycare centers in Afghanistan" (Larson, 2014). Small wonder, then, that the Association for Unmanned Vehicle Systems International (AUVSI), the industry's leading trade organization, launched a campaign to persuade journalists to stop using the word "drone" when writing stories about their airborne enterprises. To that end, reporters covering the 2013 AUVSI convention in Washington, DC discovered that the Wi-Fi password in the press room was "DONTSAYDRONE."

Relishing the irony, Brian Fung (2013) responded with a blog post headlined "Why Drone Makers Have Declared War on the Word 'Drone.'" Some joke. For the hundreds, perhaps thousands of civilian casualties of America's drone wars,[1] these high-tech killing machines may not be four-letter words, but they most certainly are an obscenity. And yet, the stories of victims and survivors of US drone strikes are rarely heard outside of independent media and the alternative press. What's more, corporate news media routinely ignore the testimony of former drone operators turned whistleblowers, as well as direct action campaigns waged against drone warfare. Nonetheless, these stories are just as relevant as news accounts of drone strikes across the Islamic world, marketing and advertising copy promoting the latest UAS, and press releases issued by business leaders and local governments extolling the economic and security benefits of UAVs. Indeed, the stories of survivors, witnesses, and resisters are essential if we are to fully comprehend the social, political, and cultural implications of drones.

This book examines the relationship between technology and culture. It starts from the premise that technology shapes and is shaped by the stories we tell about it. As communication scholar James Carey observed:

> Stories about technology … play a distinctive role in our understanding of ourselves and our common history. Technology, the hardest of material artifacts, is thoroughly cultural from the outset: an expression and creation of the very outlooks and aspirations we pretend it merely demonstrates. (Carey, 1992, p. 9)

Drawing on insights from media, communication, and technology studies, this book explores how news accounts and op-eds, marketing and advertising, as well as all manner of popular culture construct public knowledge of drones. Throughout, I argue that media discourse plays a decisive role in *shaping* these new technologies, *understanding* their application in various spheres of human activity, and *integrating* them into everyday life. Furthermore, I contend "drone discourse" employs a common trope associated with scientific and technical innovation: the technological sublime (Marx, 1964; Nye, 1994). That is to say, media narratives about drones—at once anxious and hopeful, fearful and awe-inspired—are emblematic of the profound cultural ambivalence that frequently accompanies the introduction of new technologies (Marvin, 1988; Rothstein, 2015). Thus, the book highlights the social construction of technology through critical analysis of the relationship between discursive and material practice.

This introduction proceeds with an analytical framework for examining this relationship. It locates this study in terms of the decisive role media discourse plays in the construction of public knowledge and the politics of culture. Following this, I describe the book's organization into four thematic sections and offer a rationale for the selection of the specific cases, episodes, and texts examined throughout. As we shall see, critical analysis of a variety of cultural forms—from newspaper headlines, nightly newscasts, and documentary films to advertising, entertainment media, and graphic arts—vividly demonstrates the prevalence of drones in global battlefields and domestic airspace, public discourse and the popular imagination.

Drone Discourse: Power, Hegemony and Resistance

From the outset, it is important to acknowledge that there is no single approach to the study of media discourse. As contributors to Bell and Garrett's (1998) seminal volume demonstrate, scholars employ a range of analytical frameworks to examine, for instance, opinion and ideology in the press (van Dijk), televisual news discourse (Allan), newspaper design and layout (Kress & van Leeuwen), and language in broadcast media (Scannell). Despite the variety of texts scholars examine, and the diverse approaches they employ, analysts share a common interest in exploring the decisive, but often opaque relationship between mediated discourse and social structures, relations, and interaction. "Media discourse is important," Bell contends, "both for what it reveals about a society and because it also itself contributes to the character of society" (1998, pp. 64–65). As we shall see, drone discourse exposes, and frequently conceals, a great deal about how, why, and to what ends drone technology has developed historically. Doing so, it reveals profound contradictions in US culture and society—especially America's "imperial ambitions" in the aftermath of the attacks of September 11, 2001 (Chomsky & Barsamian, 2005).

Furthermore, drone discourse draws upon and recirculates a wellspring of cultural attitudes and sensibilities toward technological innovation more generally (Rothstein). In other words, media representations of drones—the words, sounds, and images that we use, and the stories that we tell—reflect as well as refract popular attitudes toward this new technology. This is not to suggest, however, that drone discourse is unique in this regard: a point taken up in greater detail in Chapter 1 with respect to the technological sublime. For present purposes, then, we begin with a concise review of the philosophical underpinnings of discourse theory. Following this, I describe the analytical approach employed in this study: critical discourse analysis (CDA). Throughout, I make every effort to avoid too much technical jargon in the hope that this book may appeal to non-specialist and lay audiences who might be unfamiliar with the sometimes esoteric language of critical discourse analysis (Billig, 2008).

Discourse Theory: Making Meaning, Constructing Knowledge

Notwithstanding its application across the humanities and social sciences, from literary studies and social linguistics to narrative theory and discursive psychology, discourse analysis is fundamentally concerned with "language in use" (Smith & Bell, 2007, p. 79). How scholars across these disparate fields define "discourse" is another matter altogether. Drawing on Michel Foucault's influential work, Charlotte Epstein defines discourse as "a cohesive ensemble of ideas, concepts, and categorizations about a specific object that frame that object in a certain way and, therefore, delimit the possibilities for action in relation to it" (2008, p. 2). Adopting a more abstract definition, Eoin Devereux refers to discourse as "a form of knowledge" (2003, p. 158). Striking a balance between these two descriptions, Marianne Jorgensen and Louise Phillips offer a definition that is parsimonious as it is incisive: "a particular way of talking about and understanding the world (or an aspect of the world)" (2002, p. 1). With its emphasis on the *sensemaking* dimensions of discourse, Jorgensen and Phillips' definition is well suited to this project. Indeed, discourse analysis is especially useful for examining the "production and circulation of meaning through language": a principal concern of contemporary media and cultural studies scholarship (Hall, 1997, p. 1).

Discourse theory starts from a basic assumption regarding the relationship between language and reality. Operating under the rubric of *social constructionism* (Berger & Luckmann, 1966; Burr, 2015), discourse theory holds that language in not a reflection of reality; rather, language plays a decisive role in constituting "the real." This is not to suggest that naturally occurring phenomenon, like thunderstorms and wildlife, are figments; nor that material artifacts, such as cell phones and airplanes, do not exist. They certainly do. However, our perception and understanding of these objects is mediated through language. As Carey puts it, "Reality is not given, not humanly existent, independent of language and toward which language stands as a pale refraction. Rather, reality is brought into existence, is produced, by communication—by, in short, the construction, apprehension, and utilization of symbolic forms" (1992, p. 25). Our "access to reality" (Jorgensen & Phillips, p. 5) is, therefore, a product of symbolic forms, like language; and discursive strategies, such as categorization, by which we classify and distinguish, for example, the natural world from the realm of material artifacts. The key insight here is that the meaning of, say, a bird or a drone, is not inherent in the objects themselves. Instead, we *ascribe meaning* to material reality through discourse.

Additionally, discourse theory holds that symbolic forms and representational systems are historically and culturally specific. For example, we can detect the evolution of the word drone from the middle of the twentieth century to the present day. During WWII, the armed forces used target drones to train pilots and anti-aircraft personnel. At the time, the word "drone" connoted an expendable, inexpensive radio-controlled (RC) aircraft; a far cry from the sophisticated, semi-automated killing machines used in US counterterrorism operations today. Over the course of the past half-century, then, what we mean when we use the word "drone" has changed dramatically. Thus, meaning is never completely fixed, but is historically contingent. What's more, the process by which material artifacts, like drones for instance, come to be known and understood is culturally specific. In the United States, drones are routinely portrayed as an effective means of combatting terrorism in a safe (for the pilots), efficient, and cost effective manner. Consequently, public opinion polls routinely find a majority of Americans support the targeted killing program, even if a terror suspect is a US citizen.[2] Across the Islamic world, however, drones are perceived quite differently.

> For a Muslim tribesman, this manner of combat not only was dishonorable but also smacked of sacrilege. By appropriating the power of God through the drone, in its capacity to see and not be seen and deliver death without warning, trial, or judgment, Americans were by definition blasphemous. (Ahmed, 2013, p. 2)

What's more, flying unseen, but not unheard, for hours and days at a time, drones inflict emotional and psychological trauma on civilian populations (Benjamin, 2015). For communities living under drones, these symbols of American technological ingenuity are a source of sheer terror (Cavallaro, Sonnenberg, & Knuckey, 2012).

Discourse theory further conceptualizes knowledge as a product of social practices, such as formal education, everyday conversation, and most important for our purposes, the daily output of the media and cultural industries. As Roger Silverstone observes, media are central to "our capacity or incapacity to make sense of the world in which we live" (1999, p. ix). Accordingly, this book's analysis of media discourse reflects the centrality of factual as well as fictional media in the construction of public knowledge. Furthermore, as Jorgensen and Phillips note: "Knowledge is created through social interaction in which we construct common truths and *compete about what is true and false*" (Emphasis added, 2002, p. 5). Put differently, knowledge is produced, circulated—and frequently contested—through social interaction. Nowhere is this more evident than in ongoing debates over the "accuracy" and "precision" of US drone strikes, let alone widely disputed estimates of civilian casualties in America's drone wars (Shane, 2011). Hence, the concept "discursive struggle" (Laclau & Mouffe, 1985), herein understood as competing ways of talking about or understanding the world, is of paramount importance to discourse analysis. In the pages that follow, we will explore this discursive struggle as it plays out between various players and interests: from military officials and anti-war activists to government regulators and drone entrepreneurs.

Finally, discourse theory examines the link between knowledge and social action. Here, the concept of ideology takes center stage. That is, if we define ideology as "an interest linked perspective," as Greg Philo suggests, ideological discourse is best understood as "a language or system of representation that has developed socially in order to make and circulate a coherent set of meanings, which serve the interests of a section of society" (Fiske, quoted in Philo, 2007, p. 101). Rarely has this interest linked perspective been made clearer than when former CIA director Leon Panetta asserted that America's drone campaign was "the only game in town" for achieving US counterterrorism objectives (CNN, 2009). Informed by a worldview that legitimates the United States' routine violation of international law, Panetta's statement naturalizes America's targeted killing program while simultaneously precluding non-lethal, non-military approaches to terrorist threats—imminent or imagined.

In sum, the premise of discourse theory is that systems of representation—spoken and written language, image and sound, as well as multimodal forms like television news discourse—are neither neutral, let alone accurate reflections of the people, objects, and events of the world. Rather, discourse plays a central and decisive role in constructing our world. The work of the analyst, therefore, is to interrogate how "the ascription of meaning in discourses works to constitute and change the world" (Jorgensen & Phillips, p. 9).

One of many approaches to this task, critical discourse analysis (CDA) has emerged as a leading framework in contemporary media studies scholarship

(Bell & Garrett). With its attention to the interplay between discursive and non-discursive practices, CDA sharpens our analytical focus on the mutually constitutive relationship between media discourse and technological artifacts. Equally important, CDA aims not only to understand social change, but also contribute to such change. "What makes critical discourse analysis 'critical' is its normative commitment to positive social change" (Jackson, 2005, p. 25). As such, my analysis of the discourse surrounding both armed and surveillance drones is not politically neutral. Rather, informed by CDA, my critique underscores, challenges, and resists the unequal relations of power embedded in drone discourse. It is to this approach to discourse analysis that we now turn.

Critical Discourse Analysis: Dominance and Change

While van Dijk (1988) traces its origins from classical rhetoric through formalism and post-structuralism, CDA's most immediate roots lie in the field of critical linguistics (Philo). Taking as their starting point a functionalist concern with the role of language in the exchange of meaning, Fowler, Hodge, Kress, and Trew's (1979) critical linguistic analysis calls attention to the ideological consequences of word choice, grammar, and syntactic structure in public discourse. In subsequent work, Fowler (1991) critiques journalistic practice, specifically the distinct but related processes of *selection* and *transformation* in the construction of news stories. That is, Fowler observes the role of professional ideologies, news values, and institutional cultures in determining which events are selected as newsworthy and how these events are depicted.

Doing so, Fowler reveals the unstated assumptions and "abstract propositions" (1991, p. 2) that determine the structure of news discourse. Moreover, Fowler's analysis underscores the decisive role of language, as well as the technical requirements of a specific medium (i.e., print, photography, radio, and television), in turning raw events into news stories.[3] It is worth noting that Fowler's microanalysis of language builds upon and compliments semiotic (Hartley, 1982) and cultural studies (Hall, 2005) approaches to media discourse that likewise critique the "cultural reproduction of power relations across society" (Allan, 1998). Synthesizing these disparate forms of cultural analysis, CDA has emerged as an especially productive approach for "analysing opaque as well as transparent structural relationships of dominance, discrimination, power and control" embedded within media discourse (Wodak, 2001, p. 2).

Two of CDA's leading practitioners, Teun A. van Dijk and Norman Fairclough, have focused a great deal of their efforts on analyzing news discourse:

a crucial site for the (re)production and maintenance of dominant ideologies. To this end, their research engages three key concepts for CDA—power, history, and ideology—as they relate, for example, to racial and ethnic stereotypes (van Dijk, 1987) and political discourse (Fairclough, 1998) in news media. Nonetheless, as Fowler reminds us news discourse is not unique in this regard.

> In fact, what is being claimed about news can equally be claimed about *any* representational discourse. Anything that is said or written about the world is articulated from a particular ideological position: language is not a clear window but a refracting, structuring medium. If we can acknowledge this as a positive, productive principle, we can go on to show by analysis how it operates in texts. (1991, p. 10)

In an era marked by a burgeoning of text types across a constellation of media platforms, it is all the more important to analyze a cross-section of what I'm calling drone discourse, including news media, of course, but also advertising and promotional material, activist and independent media, as well as popular culture.

In short, this project embraces CDA's commitment to close analysis of media texts for purposes of cultural and ideological critique. Hence the "wide net" cast over all manner of media texts throughout these pages. Likewise, this project shares CDA's refusal to examine texts in splendid isolation. That is to say, like cultural studies and social semiotics, CDA analyzes media texts themselves, as well as their production and reception. So too, this study aims to reveal the relationship between media discourse, power, and ideology through close reading of media texts in their social, cultural, and political contexts.

Specifically, this project employs Fairclough's (1995) tripartite analytical framework for CDA—by some accounts "the most developed theory and method for research in communication, culture and society" (Jorgensen & Phillips, p. 60). Over the past three decades, Fairclough has developed and refined this three-dimensional model consisting of: (1) a *text*; (2) *discursive practice*; and (3) *social practice*. Here, a text is broadly understood to include spoken and written language, sound and image or some combination thereof. The task of the analyst is to examine the formal features of a text, such as wording, grammar, metaphor, and collocation, or words that are frequently placed together. For instance, news reports routinely employ collocations such as "Muslim extremist" and "Islamic militant" to describe the targets of US drone strikes. Nominalization, or the process of transforming verbs into nouns, is another linguistic feature common to such accounts. For example, the nominal phrase "drone strike" transforms an action into an object or event: in this instance, an event without a discernable agent responsible for it.

The second dimension of Fairclough's model, discursive practice, refers to processes related to the production and reception of a text.

> Analysis of discursive practice focuses on how authors of texts draw on already existing discourses and genres to create a text, and on how receivers of texts also apply available discourses and genres in the consumption and interpretation of texts. (Jorgensen & Phillips, p. 69)

Here the related concepts of *interdiscursivity*, *intertextuality*, and *intertextual chains* are especially useful for analyzing discursive practice. For example, journalists routinely invoke Hollywood films like *RoboCop* and the *Terminator* series in news stories about recent developments in drone technology. Doing so, journalists combine news with science fiction discourse in entertaining, sometimes incisive ways. For Fairclough, this interdiscursivity is both a response to and a harbinger of social and cultural change. Without putting too fine a point on it, drones represent a dramatic change in the character and conduct of modern warfare (Chamayou, 2015). What's more, as veteran journalist Richard Whittle observes, "drones of all kinds are now poised to transform civilian aviation, law enforcement, agriculture, and dozens of other human endeavors" (2014, p. 6).

A form of interdiscursivity, intertextuality is unavoidable. Texts routinely draw on previous texts in explicit or implicit fashion: a situation that has implications for media producers and audiences alike. For instance, when news workers draw on popular culture in the fashion described above, they cue audiences to interpret drones and related technologies through the lens of dystopian imagery and themes. Thus, intertextuality is central to the production and circulation of meaning. Likewise, intertextual chains are useful for charting how texts circulate and change over time. Consider Amazon CEO Jeff Bezos's December 1, 2013 appearance on CBS television's long-running news magazine *60 Minutes*. With great fanfare, Bezos announced the online retailer's futuristic plan to deliver packages by drone. This "news story"[4] has since been referenced in scores of reports about the Federal Aviation Administration's (FAA) regulatory approach to drones operating in civilian airspace. So too, rivals and competitors quickly took to the airwaves to promote their wares by poking fun at Amazon's scheme. What's more, Amazon's announcement has been the subject of dozens of internet parody ads. In short, intertextual chains reveal how professional producers and popular audiences alike transform the content, style, and stance of media texts.

Social practice, the third element in Fairclough's model, indicates the broader context in which texts are produced and consumed. Fairclough identifies two aspects to this context. The first is referred to as "orders of discourse" or the network of discourses to which a particular discursive practice belongs—a point we return to in the chapter outline below. The second aspect considers the non-discursive forces that structure discursive practice. Here, Fairclough situates discursive practice in relation to broader political, economic, and institutional factors and conditions. For instance, analyzing the aforementioned CBS news story on Amazon's plans for drone delivery would take into account the mutually dependent relationship between corporate media organizations and commercial advertisers. CBS promoted Bezos's appearance in an effort to enhance ratings for that evening's broadcast. Conversely, the news magazine's report served as free publicity for the online retailer. To fully comprehend the meaning of the *60 Minutes* story, therefore, requires looking beyond the text itself, or the discursive practices associated with its production and consumption. Rather, analysts must also consider non-discursive elements, including social, political, and economic structures and relationships: what Fairclough refers to as "the social matrix of discourse" (1992, p. 237). As we shall see, social practice is essential for understanding the news discourse produced by mainstream news organizations, as well as the work of independent journalists and activists working outside of corporate media. Likewise, any evaluation of marketing campaigns aimed at the domestic drone market must be understood in terms of the social matrix of discourse associated with the advertising and public relations industries.

To summarize, Fairclough's model provides a theoretically sophisticated approach to empirical analysis of media discourse. Doing so, this approach highlights the role of discourse in (re)producing, challenging, and changing public knowledge and social relations, including unequal relations of power. As I aim to demonstrate in the following pages, drones have altered the conduct of twenty-first century war. And with the advent of drones operating in domestic airspace, this new technology has wide-ranging implications for law enforcement, surveillance, and privacy, as well as commerce, leisure, and recreation (Rothstein). In short, what I'm calling drone discourse has and will continue to play a decisive role in constructing public understanding of drones, and integrating these new technologies into our public and private lives.

Finally, with its attention to non-discursive forces, described above as "social practices," Fairclough's model requires scholars to blend discourse analysis with social and cultural theory, thereby lending this approach its interdisciplinary character. This study integrates discourse theory with theories of technological innovation in an effort to better understand the character, constitution, and consequences of drone discourse. Where this project departs from traditional approaches to CDA is in the breadth of media texts under consideration. Routinely, CDA involves the rigorous and systematic analysis of a discrete number of texts. This project takes a different tack: one that has been recognized as a welcome addition to more comprehensive analysis of relatively small samples. As Allan Bell argues: "Because the analyses are time-consuming if undertaken at all comprehensively, we need to complement full-scale work on a few texts with more piecemeal, specific analyses on larger samples" (1998, p. 103).

To that end, I consulted hundreds of media texts including news reports and political analyses, industry, public relations and marketing communication, and all manner of popular culture. As there is no single or preferred method of gathering data for purposes of CDA (Wodak & Meyer, 2009, p. x), I employed a "pull/push" strategy for data collection. For instance, using keyword searches of the Lexis-Nexis database, I retrieved print and broadcast media transcripts of news accounts of specific episodes in the US drone campaign—what I refer to throughout as critical discourse moments. Conversely, creating a Google alert using the word "drone" and subscribing to trade group and specialist sources, such as *sUAS News*, an industry newsletter, and the *Weekly Roundup*, a digest of news and commentary distributed by the Center for the Study of the Drone, delivered relevant material straight to my email in-box. Finally, data collection was not a discrete phase of the research process, but rather an ongoing procedure, which lead to the development and refinement of concepts and categories of analysis (Wodak & Meyer). The texts examined herein constitute a representative sample of this material with an eye toward tracing the contours of various orders of drone discourse. Attending to disparate discourses surrounding drones in this fashion, I hope to underscore the diversity of this discourse, while simultaneously highlighting the contest between different ways of talking about and understanding this new technology.

Significantly, this book not only draws upon and analyzes the variety and proliferation of drone discourse; it contributes to this discourse as well. Herein we can detect the political orientation of this project: an effort marked by a commitment to social change through scholarly inquiry and intellectual labor. On that score, this book is consistent with the traditions of discourse analysis and likeminded approaches to cultural critique. Like other instances of critical social research, then, this book "aims to uncover the role of discursive practice in the maintenance of unequal power relations, with the overall goal of harnessing the results of CDA to the struggle for radical social change" (Jorgensen & Phillips, p. 64). Put differently, this project employs discourse analysis in an effort to better understand how various publics interpret the rise of the drones, and, by adopting a normative perspective to new and emergent technologies, how best we might integrate these flying machines into our daily lives.

Chapter Outline

The book is organized into four parts. Each part corresponds to what Norman Fairclough, following Michel Foucault, describes as "orders of discourse." According to Fairclough, an order of discourse is "a social structuring of semiotic difference in a particular social ordering of relationships amongst different ways of making meaning, i.e. different discourse and genres and styles" (Fairclough, 2001, p. 2). Further, Fairclough conceptualizes orders of discourse operating within specific institutions or social domains—such as the order of discourse of the university, the health care system, or the media. As we shall see, drone discourse operates across several domains: international relations, domestic politics, and the military; business, law enforcement, and civic affairs; as well as the "ordinary talk" of anti-war activists and witnesses to US drone strikes.

Of course, there is frequent overlap between these orders of discourse when, for example, domestic law enforcement officials use the discourse of national security to justify the use of drones for surveillance and crowd control. Moreover, news discourse operates across each of these domains. Nevertheless, as Fairclough observes: "One aspect of this ordering is dominance, some ways of making meaning are dominant or mainstream in a particular order of discourse, others are marginal, or oppositional, or alternative" (Fairclough, 2001, p. 2). The book's organization reflects this ordering. That is to say, the book's structure underscores the contested and competitive character of drone discourse within and between disparate and unequal orders of discourse. In the pages that follow, I examine these orders of discourse with an eye toward identifying and analyzing the discursive structures, strategies, and practices associated with each.

Part I identifies discourse surrounding weaponized drones as the dominant order of discourse. Specifically, Chapters 1–3 examine media discourse related to the development and subsequent deployment of drones in twenty-first century warfare. To this end, Chapter 1 traces the cultural, linguistic, and technological origins of drones across a variety of discursive formations, from ancient mythology and technology exhibitions to science fiction Throughout, I highlight the rhetoric of the technological sublime at work in these discursive formations. Attending to the cultural antecedents of drones in this fashion, this chapter employs the concept of "anticipatory discourse" (Dunmire, 2007; Scollon & Scollon, 2000) to consider the role of language and narrative in imagining, articulating, and legitimating new technologies. Chapter 2 applies this concept to the discursive strategies employed by the George W. Bush administration in the early years of the targeted killing program. The chapter further identifies and analyzes news routines and practices that would come to define press coverage of the American drone campaign. Chapter 3 considers the preeminent role drones played in US counterterrorism strategy under the Obama administration. During his tenure, Barack Obama made extensive use of drones in declared war zones, such as Iraq and Afghanistan, as well as undeclared conflict areas including Pakistan, Yemen, and Somalia. Throughout, I focus on news coverage surrounding leaked documents regarding President Obama's secret "kill list" and the administration's legal justification for assassinating US citizens without charge. Doing so, I chart the trajectory of the on-again off-again public debate over targeted killing in the Obama years.

Part II examines a second of order of discourse: what might usefully be described as a discourse of domestication. This discursive formation reveals the difficulty drone enthusiasts and entrepreneurs face in overcoming the secrecy, if not the stigma, surrounding the drone assassination program. To that end, Chapter 4 explores the role of news and marketing discourse in promoting drones as consumer goods, career opportunities, and engines of economic development (Dillow, 2013). In addition, this chapter examines representations of drones in film and television, as well as this new technology's increasingly prominent role in producing popular entertainment. Chapter 5 considers the use of unmanned vehicles in all manner of electronic surveillance, from law enforcement and border security to conservation projects. In addition, this chapter explores the drone's unrivaled capacity to capture unfolding events from unusual, often dramatic aerial perspectives (Kaufman & Somaiya, 2013). To this end, the chapter considers the limits and possibilities of what practitioners have dubbed "drone journalism" (Tremayne & Clark,

2014). Throughout, this chapter examines the rhetoric of law enforcement, environmentalism, immigration reform, and journalistic innovation in news coverage of the increasingly widespread use of drones in domestic airspace. Following this, Chapter 6 examines domestic news coverage of the drone wars with special emphasis on propaganda, spin, and related "perception management techniques" (Solomon, 2005) designed to manufacture consent for America's drone wars. Doing so, this chapter reveals the "ideological effect" (Fairclough, 2003; van Dijk, 1998) of reporting the drone wars: namely, public acceptance of, if not indifference to, the pernicious logic of killing by remote control.

Part III examines yet another order of discourse: witnessing. Witnessing is a privileged form of media discourse, one common to news stories featuring first-hand accounts of dramatic or spectacular events (Wang, Lee, & Wang, 2013). In the context of America's drone wars, however, eyewitness accounts are infrequent at best. In this light, witnessing constitutes an alternative discourse to the dominant narrative surrounding weaponized drones discussed in Parts I and II. Chapter 7 examines news accounts that feature the voices and opinions of witnesses to drone strikes. While these voices are rarely heard in US corporate media, alternative and independent press accounts routinely include such perspectives. Surveying these stories, this chapter highlights the political and cultural significance of witnessing in the era of drone warfare. For instance, on April 23, 2013, Yemini activist and journalist Farea al-Muslimi testified before a US Senate subcommittee on the impact of drone warfare in his homeland. Despite his rare and compelling testimony, al-Muslimi's remarks received limited and rather circumscribed coverage in US corporate media. Chapter 8 analyzes al-Muslimi's testimony and the ensuing media blackout of his remarks. Like other chapters in this section, this discussion examines al-Muslimi's testimony in light of recent work on witnessing in contemporary media culture (Frosh & Pinchevski, 2011).

Drone operators represent another species of witness to America's targeted killing program. During combat missions, unmanned planes offer remote pilots an unprecedented view of their targets, allied forces, and the surrounding terrain (Gregory, 2011). Furthermore, the sophisticated optical and imaging technologies onboard unmanned planes yield vivid images of victims of US drone strikes—"bug splat" in military slang. Chapter 9 examines public statements made by military personnel who fly weaponized drones. For critics, drone pilots are glorified video gamers whose experience of warfare is sanitized and safely removed from the perils of battle. However, drone pilots are required to surveil their targets for days, perhaps weeks at a time, thereby engendering a decidedly unidirectional form of "distributed intimacy" (Cohen, 2010; Edmondson, 2001) between distant pilots and their prey. What's more, remote pilots view the aftermath of drone strikes in graphic detail. When they speak publicly about these attacks, drone operators "bear witness" (Tait, 2011) to the brutality of twenty-first century warfare.

Part IV explores oppositional drone discourse common to direct action campaigns, activist media, and popular culture. Throughout, I consider the efficacy of this cultural production in broadening the debate over the use of drones in US counterterrorism efforts. As resistance to drone warfare grows, this debate is happening across the United States, from the outskirts of Hancock Field Air National Guard Base in Syracuse, NY, to the floor of the US Senate (Bard Epstein, 2013). At the same time, anti-drone activism is increasingly common around the world (Benjamin). Despite this global resistance, however, direct action campaigns receive scant press coverage in US corporate media. As a result, activist groups must rely on local news coverage to publicize their efforts or produce their own media to mobilize resistance (Ryan, 1991). Chapter 10 examines local reporting on civil disobedience and other nonviolent tactics aimed at grounding the drones.

Conversely, Chapter 11 explores cultural resistance (Duncombe, 2002) to drones in contemporary graphic arts. Specifically, I analyze the visual rhetoric of a series of Photoshop images depicting slain civil rights leader Martin Luther King, Jr. and US President Barack Obama. Employing the internet meme "I Have a Drone," the images highlight the disparate political philosophies of two iconic African-American leaders. As we shall see, this meme is emblematic of the humorous, often incisive, blending of dissident politics and popular culture in digital space (Shifman, 2014). Scholars have long observed satire's ability to articulate resistant sensibilities and practices. Drawing on scholarship in contemporary political communication, Chapter 12 examines cultural resistance to drones in the satiric register (Day, 2011; Gray, Jones, & Thompson, 2009). Throughout, I analyze the discursive strategies common to editorial cartoonists, late night comedians and others who employ irony and satire as a form of resistance to weaponized drones—and their domestic counterparts.

The book's final chapter considers drones in light of Harold Innis's seminal work on empire and communication (1995, 2007). Without putting too fine a point on it, the US drone program exemplifies Innis's thinking on the limits and possibilities of time- and space-binding media. Notwithstanding criticism that his historical analysis of technological development suffers from rigid determinism, Innis understood full well the significance of language, culture, and deliberative politics in shaping technology (Carey, 1992). Only in this way can we fully reckon with drones, not as the inevitable realization of technological progress and superiority, but rather in terms of a culturally specific, material-discursive practice: the latest manifestation of the American technological sublime. Doing so, the book concludes with an appropriately discursive turn: one that seeks to understand drones by way of Innis's originality of thought and the clarity of his moral vision.

Notes

1. Due to the lack of transparency surrounding America's covert drone program, estimates of civilian casualties vary dramatically. See Serle (2016).
2. For example, see this Voice of America news story on a 2015 Associated Press/GFK public opinion poll. Available at http://www.voanews.com/content/most-americans-support-drone-strikes-on-terrorists/2744585.html.
3. Cultural theorist Stuart Hall makes much the same argument in his influential essay on audience interpretation of media discourse. See Hall, S. (2005). Encoding/Decoding. In S. Hall, D. Hobson, A. Lowe, & P. Willis (Eds.), *Culture, media, language* (pp. 117–127). New York, NY: Routledge.
4. Bezos's announcement was short on technical details and long on hype. The segment, which aired the night before "Cyber Monday," the start of the online holiday shopping season, was more publicity stunt than hard news story. Available at http://www.cbsnews.com/videos/amazon-unveils-futuristic-plan-delivery-by-drone/.

References

Ahmed, A. (2013). *The thistle and the drone: How America's war on terror became a global war on tribal Islam*. Washington, DC: Brookings Institution Press.

Allan, S. (1998). News from nowhere: Televisual news discourse and the construction of hegemony. In A. Bell & P. Garrett (Eds.), *Approaches to media discourse* (pp. 105–141). Malden, MA: Blackwell.

Bard Epstein, A. (2013). Protesting drone deaths. *The Nation*. Retrieved from http://www.thenation.com/article/174030/noted

Bell, A. (1998). The discourse structure of news stories. In A. Bell & P. Garrett (Eds.), *Approaches to media discourse* (pp. 64–104). Malden, MA: Blackwell.

Bell, A., & Garrett, P. (Eds.). (1998). *Approaches to media discourse*. Malden, MA: Blackwell.

Benjamin, M. (2015). The grim toll drones take on innocent lives. In M. Cohn (Ed.), *Drones and targeted killing: Legal, moral, and geopolitical issues* (pp. 91–98). Northampton, MA: Olive Branch Press.

Berger, P. L., & Luckmann, T. (1966). *The social construction of reality: A treatise in the sociology of knowledge*. Garden City, NY: Doubleday.

Billig, M. (2008). The language of critical discourse analysis: The case of nominalization. *Discourse & Society, 19*(6), 783–800.

Bowman, T. (2012, August 10). Air Force chief leaves legacy in the sky. *NPR*. Retrieved from http://www.npr.org/2012/08/10/158521495/air-force-chief-leaves-legacy-in-the-sky-drones

Burr, V. (2015). *Social constructivism* (3rd Edition). New York, NY: Routledge.

Carey, J. W. (1992). *Communication as culture: Essays on media and society*. New York, NY: Routledge.

Cavallaro, J. Sonnenberg, S., & Knuckey, S. (2012, September 25). Living under drones: Death, injury and trauma to civilians from US drone practices in Pakistan. *Stanford Law School*. Retrieved from https://law.stanford.edu/publications/living-under-drones-death-injury-and-trauma-to-civilians-from-us-drone-practices-in-pakistan/

Chamayou, G. (2015). *A theory of the drone*. New York, NY: The New Press.

Chomsky, N., & Barsamian, D. (2005). *Imperial ambitions: Conversations on the post-9/11 world*. New York, NY: Metropolitan Books.

CNN. (2009, May 18). US airstrikes in Pakistan called "very effective." *CNN.com*. Retrieved from http://www.cnn.com/2009/POLITICS/05/18/cia.pakistan.airstrikes/

Cohen, K. (2010). *Never alone, except for now: Mediated collectivity in networks* (Unpublished doctoral dissertation). University of Chicago. Retrieved from http:/gradworks.umi.com/3432709.pdf

Day, A. (2011). *Satire & dissent: Interventions in contemporary political debate*. Bloomington, IN: Indiana University Press.

Devereux, E. (2003). *Understanding the media*. London: Sage.

van Dijk, T. A. (1987). *Communicating racism: Ethnic prejudice in thought and talk*. Newbury Park, CA: Sage.

van Dijk, T. A. (1988). *News as discourse*. Hillsdale, NJ: Lawrence Erlbaum.

van Dijk, T. A. (1998). Opinions and ideologies in the press. In A. Bell & P. Garrett (Eds.), *Approaches to media discourse* (pp. 21–63). Malden, MA: Blackwell.

Dillow, C. (2013, May 13). Inside the drone economy. *Fortune*. Retrieved from http://fortune.com/2013/05/13/inside-the-drone-economy

Duncombe, S. (Ed.). (2002). *Cultural resistance reader*. London: Verso.

Dunmire, P. (2007). "Emerging threats" and "coming dangers": Claiming the future for preventive war. In A. Hodges & C. Nilep (Eds.), *Discourse, war and terrorism* (pp. 19–43). Amsterdam: John Benjamins.

Edmondson, R. (2001). Studying civic culture ethnographically and what it tells us about social capital: Communities in the west of Ireland. In P. Dekker & E. M. Uslaner (Eds.), *The politics of everyday life* (pp. 59–72). London: Routledge.

Epstein, C. (2008). *The power of words in international relations: Birth of an anti-whaling discourse*. Cambridge: MIT Press.

Fairclough, N. (1992). *Discourse and social change*. London: Longman.

Fairclough, N. (1995). *Media discourse*. London: Edward Arnold.

Fairclough, N. (1998). Political discourse in the media: An analytical framework. In A. Bell & P. Garrett (Eds.), *Approaches to media discourse* (pp. 142–162). Malden, MA: Blackwell.

Fairclough, N. (2001). *The dialectics of discourse*. Retrieved from www.ling.lancs.ac.uk/staff/norman/2001a.doc

Fairclough, N. (2003). *Analyzing discourse*. London: Routledge.

Fowler, R. (1991). *Language in the news: Discourse and ideology in the press*. London: Routledge.

Fowler, R., Hodge, B., Kress, G., & Trew, T. (1979). *Language and control*. London: Routledge.

Frosh, P., & Pinchevski, A. (Eds.). (2011). *Media witnessing: Testimony in the age of mass communication*. New York, NY: Palgrave Macmillan.

Fung, B. (2013, August 16). Why drone makers have declared war on the word "drone." *Washington Post*. Retrieved from https://www.washingtonpost.com/news/the-switch/wp/2013/08/16/why-drone-makers-have-declared-war-on-the-word-drone/?utm_term=.115cac58083c

Garamone, J. (2014, May 22). Military uses remotely piloted aircraft ethically. *Defense.gov*. Retrieved from http://archive.defense.gov/news/newsarticle.aspx?id=122308

Gillespie, J. (2013, June 9). It's not a drone—It's a cuddly flying spy bot. *The Sunday Times*. Retrieved from http://www.thetimes.co.uk/article/its-not-a-drone-its-a-cuddly-flying-spy-bot-jpb2cxpmfvn

Gray, J., Jones, J. P., & Thompson, E. (Eds.). (2009). *Satire TV: Politics & comedy in the post-network era*. New York, NY: New York University Press.

Gregory, D. (2011). From a view to a kill: Drones and late modern war. *Theory, Culture & Society*, *28*(7–8), 188–215.

Hall, S. (Ed.). (1997). *Representation: Cultural representations and signifying practices*. London: Sage.

Hall, S. (2005). Encoding/decoding. In S. Hall, D. Hobson, A. Lowe, & P. Willis (Eds.), *Culture, media, language: Working papers in cultural studies, 1972–79* (pp. 117–127). New York, NY: Routledge.

Hartley, J. (1982). *Understanding news*. London: Methuen.

Innis, H. A. (1995). *The bias of communication*. Toronto: University of Toronto Press.

Innis, H. A. (2007). *Empire and communications*. Toronto: Dundurn Press.

Jackson, R. (2005). *Writing the war on terrorism: Language, politics and counter-terrorism*. Manchester: Manchester University Press.

Jorgensen, M., & Phillips, L. J. (2002). *Discourse analysis as theory and method*. London: Sage.

Kaufman, L., & Somaiya, R. (2013, November 25). Drones offer journalists a wider view. *New York Times*. Retrieved from http://www.nytimes.com/2013/11/25/business/media/drones-offer-journalists-a-wider-view.html?_r=0

Kress, G., & van Leeuwen, T. (1998). Front pages: (The critical) analysis of newspaper layout. In A. Bell & P. Garrett (Eds.), *Approaches to media discourse* (pp. 186–219). Malden, MA: Blackwell.

Laclau, E., & Mouffe, C. (1985). *Hegemony and socialist strategy: Towards a radical democratic politics*. London: Verso.

Larson, T. (2014, February 5). We don't call them drones anymore. *San Diego Reader*. Retrieved from http://www.sandiegoreader.com/news/2014/feb/05/cover-we-dont-call-them-drones-anymore/#

Marvin, C. (1988). *When old technologies were new: Thinking about electric communication in the nineteenth century*. New York, NY: Oxford University Press.

Marx, L. (1964). *The machine in the garden: Technology and the pastoral ideal in America.* New York, NY: Oxford University Press.

Nye, D. E. (1994). *American technological sublime.* Cambridge: MIT Press.

Pappalardo, J. (2013, December 3). Drone is not a four-letter word. *Popular Mechanics.* Retrieved from http://www.popularmechanics.com/military/a9776/drone-is-not-a-four-letter-word-16228330/

Philo, G. (2007). News content studies, media group methods and discourse analysis: A comparison of approaches. In E. Devereux (Ed.), *Media studies: Key issues and debates* (pp. 101–133). London: Sage.

Rothstein, A. (2015). *Drone.* New York, NY: Bloomsbury Publishing.

Ryan, C. (1991). *Prime time activism: Strategies for grassroots organizing.* Boston, MA: South End Press.

Scannell, P. (1998). Media—language—world. In A. Bell & P. Garrett (Eds.), *Approaches to media discourse* (pp. 251–267). Malden, MA: Blackwell.

Scollon, S., & Scollon, R. (2000). *The construction of agency and action in anticipatory discourse: Positioning ourselves against neo-liberalism.* Paper presented at the Third Conference for Sociocultural Research UNICAMP, Sao Paulo, Brazil.

Serle, J. (2016, July 1). Obama casualty numbers a fraction of those recorded by the Bureau. *The Bureau of Investigative Journalism.* Retrieved from https://www.thebureauinvestigates.com/2016/07/01/obama-drone-casualty-numbers-fraction-recorded-bureau/

Shane, S. (2011, August 11). CIA claim of no civilian deaths from drones is disputed. *New York Times.* Retrieved from http://www.nytimes.com/2011/08/12/world/asia/12drones.html

Shifman, L. (2014). *Memes in digital culture.* Cambridge: MIT Press.

Silverstone, R. (1999). *Why study the media?* London: Sage.

Smith, P., & Bell, A. (2007). Unraveling the web of discourse analysis. In E. Devereux (Ed.), *Media studies: Key issues and debates* (pp. 78–100). London: Sage.

Solomon, N. (2005). *War made easy: How presidents and pundits keep spinning us to death.* Hoboken, NJ: John Wiley & Sons.

Tait, S. (2011). Bearing witness, journalism and moral responsibility. *Media, Culture & Society,* 33(8), 1220–1235.

Tremayne, M., & Clark, A. (2014). New perspectives from the sky. *Digital Journalism,* 2(2), 232–246. doi:10.1080/21670811.2013.805039

Wang, B. Y., Lee, F. L., & Wang, H. (2013). Technological practices, news production processes and journalistic witnessing. *Journalism Studies,* 14(4), 491–506.

Whittle, R. (2014). *Predator: The secret origins of the drone revolution.* New York, NY: Henry Holt.

Wodak, R. (2001). What CDA is about—A summary of its history, important concepts and its developments. In R. Wodak & M. Meyer (Eds.), *Methods of critical discourse analysis* (pp. 1–14). London: Sage.

Wodak, R., & Meyer, M. (2009). *Methods of critical discourse analysis* (2nd ed.). London: Sage.

PART I

PERPETUAL WAR

· 1 ·

TECHNOLOGICAL DREAMS AND KILLING MACHINES, OR DRONES AND THE SUBLIME

If we continue to develop our technology without wisdom or prudence, our servant may prove to be our executioner.

—US General Omar Bradley, Armistice Day Speech, November 11, 1948

A group of workers assembled in the lobby of a sleek office building watch anxiously as a swarm of quadcopters, some carrying shipping boxes, hover outside. "I think we can make it," says a shaken executive. "Right?" Our hero steps forward and instructs his colleagues: "Just stay calm and move as quietly as possible." Once outside, however, the office workers panic and the delivery drones commence their attack. While his co-workers succumb to the drones' unnerving assault, our hero retreats to his Audi A6. Thanks to his cool head and quick thinking—not to mention Audi's voice recognition software and on-board navigation system—he makes his escape by out maneuvering the pursuing drones. The advertisement's tag line reads: "Technology doesn't have to be intimidating." With drones substituting for a murder of crows that attack a rural schoolhouse, this television commercial makes unmistakable reference to Alfred Hitchcock's 1963 thriller *The Birds*.[1] No less for discourse analysts than for film buffs, the 60-second spot, which debuted in March 2015 during high-profile cable and network prime time slots, is striking.

As noted in the introduction, critical analysis of discursive practice—that is, how media texts are produced and consumed—routinely focuses on interdiscursivity, intertextuality, and intertextual chains. Audi's *The Drones* advertisement displays each of these elements of discursive practice. First and foremost, the ad exhibits a high degree of interdiscursivity. Blending the promotional discourse of television advertising with the uncanny and fantastic discourse of Hitchcock's late masterpiece, the commercial employs several narrative strategies—irony, suspense, and spectacle—to celebrate technological innovation in the wake of mechanized mayhem. Furthermore, the advertisement's skeptical attitude toward drones draws from a wellspring of cultural anxieties prevalent in contemporary media discourse, from fears of invasion of privacy to misgivings about autonomous technology (Dzieza, 2013). Second, the Audi ad makes explicit intertextual reference to *The Birds*. Nowhere is this more evident than when a young woman flees the drone assault by locking herself into a parking booth; just as Hitchcock's heroine finds relative safety from attacking gulls in a phone booth. In the commercial, a drone, rather than a bird, crashes into the window in terrifying, single-minded pursuit of its prey. Finally, with images of maniacal quadcopters carrying parcels, *The Drones* is a discursive link in an intertextual chain that runs through any number of parodies based on Amazon's advertising campaign for Prime Air, the online retailer's highly publicized same-day drone delivery service.[2]

Simply put, *The Drones* draws upon and rearticulates disparate, sometimes fantastic discourses on nature, technology, and culture. For the car manufacturer, the meaning is clear: drones may be innovative, futuristic perhaps, but these flying machines are strangely terrifying. Besides, they are no match for the computing and engineering prowess of the new Audi A6. More importantly, the advertisement reflects deep-seated, often conflicted and contradictory attitudes toward machines, and technological innovation more generally. Viewed in this light, *The Drones* is emblematic of what Leo Marx (1964), in his groundbreaking cultural analysis *The Machine in the Garden*, describes as the rhetoric of the technological sublime. Put differently, the commercial is not simply a clever bit of advertising that satirizes Amazon's marketing campaign. Nor is it unique in its ambivalence toward technological innovation. Rather, like many of the media texts examined in this book, the advertisement is best understood as an expression of a long history of public encounters with new technologies: sublime experiences at once awe-inspiring and terrifying, reverential and apprehensive.

This chapter proceeds with a discussion of the technological sublime—as David Nye (1994) observes, a "preferred trope" in American cultural history. Following this, I examine the role of anticipatory discourse—from ancient mythology and technology exhibitions to science fiction—in shaping technological innovation, capturing the public imagination, and integrating new technologies into daily life. The chapter concludes with some brief observations on the prevalence of drones in contemporary public discourse. Throughout, I demonstrate that the technological sublime is not inherent in the material artifact itself; rather, as Nye argues, it is a discursive construct—one that "articulates a distinct social and political relation to technology" (p. xiv).

The Sublime

In the Western tradition, the sublime has its roots in philosophy, aesthetics, and literary criticism. Whether it is ascribed to the natural world or the product of human intellect and creativity, the sublime refers to a sense of wonder or amazement. Furthermore, the sublime imbues natural sites and human achievement alike with spiritual and moral value—frequently evoking ecstatic, transcendent, sometimes religious states of being. Fittingly, for my purposes here, the word's archaic definition, taken from the Latin, literally means "high up in the air." Hence notions of grandeur and loftiness are commonly associated with the sublime. Equally important, however, the sublime conveys feelings of negative lust.[3] In other words, sublime objects regularly evoke powerful feelings of *attraction and repulsion*. David Nye captures these contradictory reactions when he refers to the sublime as "repeated experiences of awe and wonder, often tinged with an element of terror, which people have had when confronted with particular natural sites, architectural forms, and technological achievements" (p. xvi).

Popularized by cultural historian Leo Marx, the concept of the technological sublime is relevant to critical analysis of drone discourse for two reasons: (1) the sublime underscores the *symbolic power* of machines; and (2) the sublime informs the *cultural politics* surrounding the introduction of new technologies. To this first point, Marx argues that in the mid-1800s the machine emerged as a potent symbol in the American imagination. Where Thomas Jefferson's vision of an agrarian democracy once held sway, machines were now cornerstones of an ascendant "official America ideology of industrialism" (p. 181) promising nothing less than national unity, social equality, and human progress on a grand scale. Marx explores the rhetorical equivalence between technological innovation and national identity evident in American letters, political speeches and, most energetically perhaps, the boosterism of an emerging industrial class. His attention to the symbolic power and cultural resonance of the machine is especially instructive for illuminating the correspondence between the technological sublime and notions of national as well as human progress.

For instance, Marx notes that as material artifacts machines provide audible and visible evidence of progress. From the mechanization of farm labor to the completion of the intercontinental railroad, machines would transform the nature of work, and unite a vast and widely dispersed nation. In this way, sublime technologies were potent material and symbolic expressions of America's Manifest Destiny—the belief that the United States was destined to spread a new, egalitarian, and prosperous civilization across the whole of North America. What's more, as mechanization and industrialization transformed the continent, politicians, manufacturers, clergymen, and business leaders habitually imbued sublime technologies with values that expressed and shaped American idea(l)s of freedom and equality. For instance, in public discourse of the time, grand infrastructure projects, such as the Erie Canal, were understood as "components of a 'moral machine.' They ensured not only prosperity but also democracy and moral health of the nation" (Nye, p. 39). Thus, the sublime articulates a relationship between technology, national identity, and moral as well as economic development. In this way, the sublime embodies "a quasi-religious faith in the power of new technologies to overcome social and material restraints" (Streeter, 1997, p. 225). As we shall see, this line of thinking informs contemporary discourse on drones in American commercial, business, and military affairs.

Conversely, the machine represented universal progress in terms of humankind's conquest of nature. Central to this idea were the twin revolutions in transportation and communication ushered in by the "dynamic sublime" of the railroad and the "electrical sublime" of the telegraph respectively.[4] In the reverential parlance of the day, these sublime technologies would forever liberate humanity from the bounds of space and time. Marx notes: "No stock phrase in the entire lexicon appears more often than the 'annihilation of space and time' … The extravagance of this sentiment apparently is felt to match the sublimity of technological progress" (p. 194). In short, mechanization represented far more than American enterprise, skill, and ingenuity; the machine further symbolized the triumph of human intellect in controlling the forces of nature and shaping the course of human history. No doubt this last point accounts for the sense of urgency and possibility that pervades the rhetoric of the technological sublime (Streeter). Small wonder, then, that for advocates and critics alike, drones prompt urgent calls for legislative and regulatory action in such diverse realms as aviation and entertainment, commerce and public safety, as well as domestic and international law (Cole, 2013; McNeal, 2012).

Marx further reveals how opposing attitudes toward technological innovation were articulated throughout public culture of the time: from popular literature and ceremonial speeches to press accounts and philosophical tracts. For Marx, competing discourses over the wisdom and desirability of supplanting the pastoral ideal with an emerging techno-capitalist system represent "a case example in the politics of American culture" (p. 217). Marx is particularly interested in exploring how nascent industrialists reconciled their faith in technology with pastoralism's deep-seated skepticism toward machines. To that end, enthusiasts wedded the idea of mechanization with pastoralism's belief in the redemptive character of the New World. To borrow Marx's phrase, this approach sought "to make a place for it [the machine] in the myth of garden" (p. 151). As such, this rhetorical strategy constituted an anticipatory discourse, a narrative that places the virtuous machine—made so by contact with the redemptive and exceptional qualities of the new republic—at the center of an American ideology of equality, prosperity, and progress.

While proponents did their best to gloss over contradictions inherent in this perspective, opponents made clear their resistance to various iterations of the technological sublime. This is not to say that most observers didn't appreciate the benefits of mechanization, particularly its capacity to eliminate the drudgery of dirty and often dangerous work: a major selling point for drone advocates in the military, law enforcement, border security, and all manner of surveillance regimes (Diab, 2014). But for critics, the costs to human dignity and moral virtue associated with technological development were too high. Accordingly, the "men-will-become-machines" trope (Marx, p. 184) employed by opponents of mechanization expressed concerns over the loss of moral certitude in an emerging industrial society. This same sentiment informs current debates over the legal, ethical, and moral consequences of killing by remote control associated with drone warfare (Cohn, 2015).

Further, this sensibility reflected fears that the machine would destroy the harmonious lifestyle envisioned by the pastoral ideal. For instance, James Carey singles out America's most renowned satirist, Mark Twain's *Connecticut*

Yankee in King Arthur's Court, for its discursive strategy of "inverting the technological sublime and portraying the new technology as a specter of disaster" (1992, p. 124). So pervasive was this attitude that Marx claims:

> One might easily cull an impressive exhibit of passages displaying hostility to the new technology (a kind of composite image of the menacing machine to set beside the sublime machine) from the periodicals of the day. But a merely quantitative sample of periodical literature would be misleading. By and large the negative response appears in … organs of small organized minorities. (p. 219)

Thus, new technologies were either celebrated or condemned, depending upon the speaker's perspective on—and social, economic, and political relationship to—an emerging industrial order. Without stating it as such, then, Marx identifies two orders of discourse surrounding sublime technologies in the nineteenth century: the dominant discourse articulated by politicians and an increasingly influential industrial class pursuing interrelated interests in westward expansion and capital accumulation; and an oppositional discourse wary of the disharmony and alienation wrought by industrialization, and determined to preserve the pastoral ideal. Doing so, Marx reveals the discursive struggle at work in the rhetoric of the technological sublime—a struggle evident in contemporary discourse surrounding the latest sublime technology to seize the public imagination: drones.

In short, enthusiasm for and resistance to new technology are expressions of what James Carey describes as "opposite poles" of the sublime (p. 207). On the one hand, sublime technologies embody ideals of freedom, equality, and morality; on the other hand, these same technologies portend tyranny, repression, and corruption. All of which supports David Nye's contention that the sublime "taps into fundamental hopes and fears" (p. xiii). Before we consider contemporary manifestations of the sublime in drone discourse, let's explore this dynamic in linguistic, discursive, and technological precursors of UAVs and related technologies.[5]

Cultural Antecedents: The Power of Anticipatory Discourse

A primary distinction between the natural and technological sublime is their relationship to temporality. Whereas the natural sublime, such as Niagara Falls or the Grand Canyon, evokes the timeless and the eternal, the technological sublime, as manifest by the space and time annihilating properties of the railroad or the rocket, "aims at the future" (Nye, p. 61). No doubt this aspect of the technological sublime accounts for the ephemeral quality of awe-inspiring technologies. From the electric light to passenger flight, yesterday's technological wonder is today's commonplace convenience. Nevertheless, the future orientation associated with the technological sublime is a key feature of its attraction—and its apprehension. Put differently, the technological sublime offers a glimpse of potential futures: some full of promise and possibility, others dreadful and dismaying. In either case, the rhetoric of the technological sublime doesn't simply announce the future; it stakes a claim on what *will be* at some future point in time. Accordingly, the future orientation of the technological sublime requires attention to the relationship between discourse and action: a relationship Scollon and Scollon (2000) characterize as "anticipatory discourse."

This concept's relevance to a discursive analysis of drones is readily apparent. Citing Foucault's work on the productive capacity of discourse, Patricia

Figure 1.1: President John F. Kennedy's goal of putting a man on the moon is emblematic of anticipatory discourse. *Photo credit: NASA. Source: Flickr.*

Dunmire argues that the "political and material significance" of discourse stems from its "function to prophesize the future" (2007, p. 22). Consider, for example, President John F. Kennedy's (1961) address to Congress, when he declared that the United States "should commit itself to achieving the goal, before this decade is out, of landing a man on the moon and returning him safely to the earth." Although Kennedy did not live to see it, his words played a pivotal role in realizing this historic achievement. None of which is to dismiss non-discursive factors—principally Cold War hostilities that prompted the space race between the United States and the Soviet Union—in fulfilling the slain president's vision (Figure 1.1). Indeed, the sublime character of the space program proved decisive in legitimating the development of the nation's Intercontinental Ballistic Missile (ICBM) arsenal for a wary American public (Nye). Nevertheless, Kennedy's remarks "not only announced what was going to happen but helped to make it happen, carrying men's minds along with it and thus weaving itself into the fabric of destiny" (Foucault, 1984, p. 112). Such is the power of anticipatory discourse.

While Dunmire and likeminded analysts typically examine political discourse to reveal how powerful groups and institutions employ discursive strategies to announce, and thereby lay claim to, some inevitable future, we can examine this dynamic across other discursive formations as well. Recalling the sublime's purchase on fundamental hopes and fears, ancient mythology offers a productive, if somewhat surprising starting point to consider discourse that anticipates today's drone technology. Put differently, if we acknowledge that "myth can signify either retrospective contemplation or future anticipation" (Deckard, 2010, p. 4), then mythology represents a cross-cultural repository for human aspirations and anxieties. As researchers at the Center for the Study of the Drone argue:

> Numerous traditions, from the Philippines to Ancient Persia, feature flying creatures that are either fearsome, invisible, or both. The drone is the metallic embodiment of this genus of mythological aerial creature that humans have spent several thousands years imagining with a combination of fear, awe, and worship. (Gettinger et al., 2014, p. 14)

The natural world frequently provided inspiration for mythic tales of ominous flying creatures. Typically, such beings were fantastic amalgamations of real animals, transformed by human imagination into terrifying winged beasts that did the bidding of gods and demons. The harpies of Greek mythology—half woman, half avian predator—are perhaps the best-known example of such sky-dwelling monsters. Small wonder, then, that Waziri tribesmen refer to drones as *ababeel*: a mythic bird described in the Quran as having defended the holy city of Mecca by hurling stones at an invading army (Shah, 2012). Similarly, today's drone engineers and designers emulate all manner of airborne creatures—from humming birds to dragon flies—in a timeless quest to master the sky. As Mark Micire, Program Director of the Defense Advanced Projects Agency (DARPA) told CNN: "Birds of prey and flying insects exhibit the kinds of capabilities we want for small UAVs. Goshawks, for example, can fly very fast through a dense forest without smacking into a tree." Micire continued, "Many insects, too, can dart and hover with incredible speed and precision" (Shadbolt, 2015).

In addition to fearsome creatures, ancient mythology imbued prehistoric technologies, such as swords, axes, and hammers, with fantastic capabilities. In Norse mythology, for instance, Freyr's Sword is a supernatural rapier capable of fighting on its own accord—a magical technology that anticipates contemporary research and development of unmanned and autonomous weapon systems. Likewise, the Norse god Thor wielded the Mjölnir, literally "that which smashes." As legend has it, this supernatural hammer never failed to hit its target, and would always return to Thor's hand regardless of how far he threw it. So too, the mythological Sharur of ancient Sumeria was a sentient mace—a club with spikes and a metal head—capable of flying great distances and either striking opponents or simply gathering information on their whereabouts. In anticipation of the "hunter-killer" capabilities of weaponized drones, the Sharur, translated as "smasher of thousands," could communicate with whoever wielded it. Thus, ancient tales of fantastic weaponry anticipate recent developments in semi-automated killing machines, such as UAVs, capable of traveling great distances, communicating with remote operators, and striking distant opponents.

Finally, it's worth noting that drones, like earlier weapon systems, such as the Poseidon submarine and the Trident missile, sometimes take their names from ancient mythology. For instance, the United States Air Force's Gorgon Stare—named after the terrifying female creatures of Greek mythology whose gaze turned men into stone—uses state-of-the-art imaging and sensing technologies that enable soldiers "to track enemy movements, even dismounted individuals, across an area measuring four kilometers in diameter" (Thompson, 2015). Similarly, one of Israel's most celebrated UAVs, a so-called "fire-and-forget" assault vehicle, is nicknamed the Harpy. In sum, ancient tales of winged beasts and supernatural tools serve as an anticipatory discourse for the development of human flight, up to and including the guided missile, manned spacecraft, and most recently, weaponized drones.

Like the ancients, moderns frequently emulate the natural world to construct sublime objects, both real and imagined. Whereas the dynamic sublimes of the steamboat and the railroad revolutionized transportation, the electric sublimes of the telegraph and the incandescent lamp offered new possibilities for human communication. In his analysis of the electric sublime, David Nye examines the transformative power of electricity in the realms of manufacturing, advertising, and commerce. Throughout, Nye highlights the crucial role technology exhibitions played in introducing the public to the emerging science of electrification. Significantly, these exhibitions were not simply demonstrations designed to astound the masses with spectacular displays of the electric sublime. Public exhibitions, such as the 1894 Columbian Exposition in Chicago or the Pan-American Exposition at Buffalo in 1901, served as an anticipatory discourse, offering visitors a glimpse of the not too distant future. While these spectacles offered a "powerful vision" of the future "one that was colored, edited, simplified, and expanded" (Nye, p. 151), at the same time an invisible revolution was taking place in the nascent field of wireless telegraphy.

One of the more imaginative and enigmatic figures in electrical engineering, Nikola Tesla, sought to demonstrate wireless telegraphy's potential for control at a distance. During the 1898 Electric Exhibition in New York City, Tesla unveiled his latest invention: the *teleautomaton*, a first-of-its kind radio controlled vehicle. With a miniature boat placed in an indoor pool specially constructed for the demonstration, Tesla operated the vessel via remote control.[6] Astonishing as the demonstration was, Tesla's device failed to leave a lasting impression with his contemporaries. According to Tesla biographer, Margaret Cheney, the invention was simply too much for the public to comprehend at the time.

> The remarkable development to which he had carried wireless, the forerunner of modern radio, would have been quite enough: but to introduce automation simultaneously, as he did, was probably too great a leap. … he was introducing an idea for which the world would not be ready for many years. (Cheney, 2001, pp. 160–161)

Undeterred, Tesla continued to promote his ideas for "increasing human energy" (2010). Writing in *Century Magazine*, Tesla explicated the concept of telautomatics: "the art of controlling the movements and operations of distant automatons" (p. 31). A principle, he added, that was applicable on land, sea, and air. Throughout this essay, written a full three years before the Wright brothers built the first successful airplane, Tesla describes the concept's implications for the future of armed conflict. Given his extraordinary foresight, especially as it relates to the development of remotely piloted vehicles (RPVs) and autonomous technologies, Tesla's thoughts on the evolution of warfare merit revisiting here at some length.

For instance, discussing the relationship between an automaton and its remote handler, what he characterizes as the machine's "borrowed mind," Tesla writes:

> By the simple means described the knowledge, experience, judgment—the mind, so to speak—of the distant operator were embodied in that machine, which was thus enabled to move and to perform all its operations with reason and intelligence. (p. 32)

With his emphasis on what we would today describe as remote control, smart technology, and semi-autonomous action, it's no wonder contemporary observers dub Tesla "the father of unmanned vehicle technology" (Weber, 2010). Tesla's demonstration, and subsequent writing on the subject of telautomatics, underscores the significance of anticipatory discourse in the development of remote controlled flight, from relatively primitive RPVs of World War I and Cold War-era guided missiles to today's weaponized drones—and beyond (Newcome, 2004). Equally important, Tesla's enthusiasm for what he foresaw as an evolution in warfare—armed conflict conducted by a handful of individuals safely removed from the battlefield—has a familiar ring to it. Tesla writes, "There is virtually no restriction as to the amount of explosives it can carry, or as to the distance at which it can strike, and failure is almost impossible" (p. 34). Not least for its prescience, if not its hubris, Tesla's vision of remote controlled weapons further anticipates the discourse of political and military leaders who routinely hail drones for their reach, firepower, and precision.

Whereas today's policy makers, Pentagon planners, and weapons manufacturers are enamored with the prospect of endless war, Tesla's motivation was quite the opposite. Tesla imagined a time when these killing machines would help realize another ancient dream: universal peace.

> To bring on this result men must be dispensed with: machine must fight machine. … The answer is simple enough: produce a machine capable of acting as though it were part of a human being—no mere mechanical contrivance, comprising levers, screws,

> wheels, clutches, and nothing more, but a machine embodying a higher principle, which will enable it to perform its duties as though it had intelligence, experience, reason, judgment, a mind! (p. 28)

Tesla's rhetoric invokes that most cherished of American symbols: the virtuous machine. Consequently, Tesla foresaw artificial intelligence (AI) as the only viable mechanism for securing global peace and security (Franklin, 2008). But judging by the scale of global investment in drones and robotic weapon systems in recent years, Tesla grossly miscalculated humanity's desire for lasting peace; today, governments and weapons manufacturers alike are rushing to be first in a "killerbot arms race" (Sofke, 2015). By July 2015, the specter of robotic warfare was so pronounced that hundreds of AI experts, including physicist Stephen Hawking and computer scientist Stuart Russell, signed an open letter calling for a global ban on autonomous weapons (Future of Life, 2015). Here again, we can detect the power of Tesla's anticipatory discourse in shaping technological innovation—if not international relations and military strategy. Small wonder, then, that advocates promote the targeted killing program as a form of "virtuous war" (Gregory, 2011).

Of course, the ultimate irony of Tesla's vision—the chilling prospect of interminable machine wars—is discernable throughout popular culture, most notably in sci-fi blockbusters such as the *Terminator* and *Transformers* franchises. All of which supports futurist Tom Lombardo's insights on the correspondence between ancient mythology, science fiction, and future oriented thinking. "As ancient mythologies provided meaning and direction for humankind, I would suggest that science fiction … will provide the stories that will give humanity meaning and direction in the future" (Lombardo, 2006, p. 12). Thus, the realm of science fiction offers another venue to consider the power of anticipatory discourse.

Much as it has done from the start, technology plays a leading role in contemporary science fiction. Early sci-fi authors Jules Vern and H. G. Wells, whose literary efforts continue to entertain and inspire, are routinely praised for the foresight of their technological dreams. As P. W. Singer notes, "science fiction has forecast real-world technologies, as well as resulting dilemmas, with stunning accuracy" (2009, p. 156). No doubt the future oriented thinking of science fiction accounts for its enduring popularity, as well as its considerable influence—especially among scientists, engineers, and military planners.[7] For instance, US Marines currently use a wristwatch to view real-time aerial footage produced by an overhead UAV. According to Colonel James Lasswell of the Marine Corps Warfighting Lab, "We got the idea from *Dick Tracy*," the tough talking comic strip detective with a penchant for technological gadgets (Singer, p. 163). Once again, science fiction functions as "a form of narration of the future" (Lombardo, p. 12).

Through stories of future times and fantastic technologies, science fiction taps into fundamental hopes and fears: hallmarks of the technological sublime. It should come as no surprise, then, that flying machines, like drones, are mainstays of the scientific imaginary (Gettinger et al., 2014). The aerial vehicles that surveil the populace in George Orwell's dystopian novel *1984* are perhaps the most familiar cultural antecedents of today's UAVs. In a less threatening register, Amazon invokes science fiction to anticipate real-world technological innovation; the tagline for its much-anticipated Prime Air service reads, "It looks like science fiction, but it's real."

Marketers are not alone in using the language of science fiction to describe drone technology. A cursory review of press accounts reveals as much. For instance, a description of Fujitsu's prototype for a bridge inspection quad-copter begins:

> The drone looks a bit like a space station from science fiction. It consists of a central unit with four rotors, a high-definition camera and sensors including gyroscopes. On either side of this central unit are two large fiberglass wheels. (Hornyak, 2015)

Likewise, the *Business Insider's* report on the aforementioned open letter calling for a worldwide ban on robotic weapons notes: "The letter's worries about 'smart' weapons that can kill on their own sounds like a science fiction trope—humans cowering in fear of killer robots they have unknowingly created" (Del Prado, 2015). Finally, *Engadget's* story on Nikola Tesla's demonstration of his teleautomaton at the 1898 Electrical Exhibition contends:

> Tesla saw both the terrible as well as the beneficial consequences for this technology when he debuted the "remote control" and the jury is still out as to whether we'll succumb to a dystopian *Terminator*-style future or reach a peaceful stasis, where we harness the usefulness of robots and autonomous devices, and avoid the worst-case scenarios. (Turi, 2014)

In short, science fiction operates as an anticipatory discourse: one that offers a productive and provocative way of talking about and understanding drones. Consider Ian Watson's (1985) short story "Slow Birds." Written during a period of renewed Cold War tensions in the early 1980s, this story describes a distant future wherein mysterious aerial weapons patrol the skies. Watson describes the flying machines in avian terms:

> They looked a little like birds, too, though only a little. Their tubular metal bodies were rounded at the head and tapering to a finned point at the tail, with two stubby wings midway. Yet these wings could hardly have anything to do with suspending their bulk in the air; the girth of a bird was that of a horse, and its length twice that of a many lying full length. Perhaps those wings controlled orientation or trim. (p. 11)

Apart from the uncanny accuracy of this description, the passage is striking for the palpable sense of wonder—and dread—it conveys. Watson continues, "In colour they were a silvery grey; though this was only the colour of their outer skin, made of a soft metal like lead" (p. 11). Here, Watson anticipates the lightweight composite materials that give military-grade drones their unprecedented reach and endurance. The emerging portrait of the slow birds bears striking resemblance to the most recognizable aerial vehicles in contemporary visual culture: America's fleet of Predator and Reaper drones.

Equally important, Watson captures the disturbing regularity of the slow birds' comings and goings.

> After each one had appeared—over hill, down dale, in the middle of a pasture or half way along a village street—it would fly onward slowly for any length of time between an hour and a day, covering any distance between a few score yards and a full mile. And vanish again. To reappear somewhere else unpredictably: far away or close by, maybe long afterwards or maybe soon. (p. 12)

Here, then, Watson's story portends the very real and altogether unnerving experience of living under drones—a point I return to in Chapter 7. In a similar vein, the celebrated science-fiction writer Philip K. Dick anticipates the cultural anxieties evoked by today's drones with the menacing, swarming robots of his novel *Vulcan's Hammer* (2012). Like Watson, Dick likens his flying hammers to all manner of winged creatures: bats, birds, and insects.

> Madness, he thought. The ultimate horror for our paranoid culture: vicious unseen mechanical entities that flit at the edges of our vision, that can go anywhere, that are in our very midst. And there may be an unlimited number of them. One of them following each of us, like some ghastly vengeful agent of evil. Pursuing us, tracking us down, killing us one by one—but only when we get in their way. Like wasps. (p. 106)

While Dick's oeuvre expresses deep suspicion of an increasingly sinister military-industrial complex, contemporary science fiction writers enjoy a far cozier working relationship with arms manufactures and Pentagon brass. Noting the popularity and influence of Orson Scott Card's 1985 novel *Enders' Game* with US armed forces, Singer writes, "He may be a writer of fiction, but like many in the field, Card also consults for the military, speaking on such topics as 'Next Generation WMD: Anticipating the Threat'" (p. 154). Similarly, sci-fi writers Greg Bear and Arlan Andrews are members of SIGMA, a self-described "science fiction think tank" that frequently works in tandem with the Pentagon, the Department of Homeland Security and other government agencies (Hall, 2007). As Pentagon planners anticipate future wars fought with ultramodern weapons, there are clear and unambiguous structural, as well as discursive links between science fiction and the military.

***

The sublime gains currency by tapping into fundamental anxieties and aspirations. At this historical juncture, drones epitomize the American technological sublime. They are, I contend, the twenty-first century synthesis of the dynamic and electric sublimes David Nye identifies in the development of intercontinental ballistic missiles during the mid-twentieth century. But drones are not unique in this regard; rather, these flying machines are simply the latest in a series of sublime technologies to seize the public imagination. Observing Americans' preoccupation with the railroad, circa 1844, Leo Marx writes, "Stories about railroad projects, railroad accidents, railroad profits, railroad speed fill the press; the fascinating subject is taken up in songs, political speeches, and magazine articles, both factual and fictional" (p. 191). Without putting too fine a point on it, one might replace the word "railroad" with the word "drone" in this statement and immediately recognize the prevalence of these flying machines in contemporary public discourse. Briefly, then, I want to illustrate the ubiquity of drones in news media and popular culture today.

Consider the worrisome frequency of press reports describing recreational drones flying perilously close to commercial airliners. Such harrowing tales typically make headlines around the world.

> **Alert on toy drones after one nearly CRASHES into Airbus at Heathrow**
>
> (*Daily Express*, December 8, 2014)
>
> **Lufthansa flight has near-miss with drone near Warsaw**
>
> (*Reuters*, June 21, 2015)
>
> **FAA: Drones flew near 4 planes in 3 days in NY, NJ**
>
> (*NJ.com*, August 3, 2015)

Each of these reports engages in a degree of press hysteria—the all caps "crashes" in the *Daily Express* headline speaks for itself. Nevertheless, it would be a mistake to underestimate the very real dangers and profound anxieties surrounding public safety expressed in these stories. Press accounts of this kind routinely follow a narrative structure that emphasizes the drama of a close encounter between a hobbyist's quadcopter and a passenger jet. The story opens with a description of the airliner, its destination, and the approximate distance between the two aircraft at the time of the incident. Reassuring statements from local authorities describing an "ongoing investigation," along with some mention of growing commercial and recreational interest in drones, follow. Such stories typically conclude with an assessment of the current state of air traffic regulations in the relevant jurisdiction. That we can clearly discern a narrative template at work in these news reports indicates the degree to which tales of near misses between jetliners and drones have become commonplace and routinized.

A palpable sense of unease informs news stories about less frequent, but no less unsettling accounts of rogue drones. For example, in March 2015, French authorities were at a loss to explain the appearance of upwards of a dozen drones flying over "sensitive areas" in and around Paris. Headlines referring to the aircraft as "mystery drones" underscore, and perhaps heighten, public anxieties. According to *Al Jazeera English*:

> In one incident in eastern Paris, witnesses saw three men retrieve the aircraft and drive off in a black car before officers could arrest them, a police source said.
>
> Earlier on Tuesday, Interior Minister Bernard Cazeneuve told parliament that about 60 drone flights over Paris and over nuclear power plants had been reported since October 5. (Al Jazeera, 2015)

With references to mysterious aircraft, black cars, and nuclear power, this story has all the makings of a political thriller. Nevertheless, the report's emphasis on danger and intrigue serves to legitimate government calls for "new legal tools" to apprehend reckless drone operators. Herein we can detect the ideological underpinnings of such stories: suspicious or illegal drone operations give legislators license to enhance and expand existing surveillance regimes—a subject taken up at length in Chapter 5. Likewise, media accounts of ominous drone activities, such as the mysterious viral video depicting a homemade quadcopter firing several rounds from a semiautomatic pistol, raised fears worldwide that would-be assassins might one day soon use consumer-grade drones to kill by remote control (RT News, 2015). The video prompted calls for additional regulations for drones operating in domestic airspace.

Then there are the growing numbers of viral videos featuring animals attacking drones. No doubt the popularity of these clips stems from humanity's propensity to anthropomorphize our hopes and fears. Consider the prevalence of so-called "bird vs. drone" videos: dramatic aerial footage—made all the more compelling with judicious editing and an appropriately thrilling soundtrack—that articulates cultural paranoia over these flying machines (Bogel, 2014). Like the anxiety-laden news stories of near misses between drones and commercial airlines, videos depicting animal aggression toward drones, such as the clip of a kangaroo punching a drone out of the sky, frequently garner global media attention.[8]

Apart from video game franchises such as *Battlefield* and *Call of Duty*, which encourage users to immerse, and thereby prove themselves in the virtual battlefields engendered by remote control warfare (Stahl, 2013), popular entertainment frequently depicts drones as loathsome and terrifying machines. With increasing frequency, drones serve as plot devices in feature films, stage plays, and television programs: a subject taken up in greater detail in Chapter 4. Even musicians have gotten into the act. The English rock band Muse's 2015 concept album, *Drones*, critiques the dehumanizing effects of robotic warfare. Conversely, hip hop artist DeWun's 2013 single "They Watching Us"—replete with intertextual references to Michael Jackson's paranoia-tinged "Somebody's Watching Me" and George Orwell's *1984*—expresses fears of government surveillance and martial law.[9]

Striking a more hopeful note, with increasing frequency news reports celebrate the benefits and versatility of drone technology. Recently, drones have made headlines for assisting search and rescue teams locate missing persons in hazardous or otherwise inaccessible areas.

> **Injured hiker rescued from Mount Finlayson, 'drone' plays a role**
> (*Times Colonist*, August 9, 2015)
> **Lifeguard drones fly to the rescue over Spanish beaches**
> (*The Times*, July 18, 2015)
> **Drones' new mission: saving lives in developing countries**
> (*The Guardian*, January 9, 2015)

The tone of these news stories reflects growing, sometimes grudging, admiration for UAVs—a welcome contrast for drone advocates eager to overcome the stigma associated with the use of drones in the US targeted killing program. Likewise, animal rights and environmental activists celebrate the drone's potential for assisting their efforts to preserve endangered species and protect fragile ecosystems. Thus, press reports play an important role in domesticating drones for humanitarian and ecological purposes. Consider the lead in this *Guardian* story about the Royal Society for the Protection of Birds (RSPB). "Military technology used to hunt down insurgents in Afghanistan has been taken up by Britain's biggest nature conservation charity to safeguard some of the country's rarest birds" (Weaver, 2014).

Other beneficial applications of drone technologies include delivering medical supplies to patients living in isolated rural communities. In July 2015, a drone made a first-of-its-kind, FAA approved delivery of pharmaceuticals to the Remote Area Medical Clinic in Wise County, Virginia. Making reference to the Wright brother's historic flight, Matthew Sweeny, co-founder of Flirtey Inc., the drone delivery start-up company leading the demonstration, told the *Wall Street Journal*, "It's absolutely a Kitty Hawk moment" (Hackman & Nicas, 2015). In short, with greater regularity, news workers produce "feel good" stories designed to offset the bad rap drones have acquired, in large part due to the secretive and controversial use of drones in US counterterrorism strategy, if not Amazon's aggressive marketing for its same-day drone delivery service (Newman, 2013).

To summarize, while drones are commonly associated with feelings of fear and terror—as the aforementioned Audi advertisement vividly, if hyperbolically, demonstrates—these devices are equally significant inasmuch as they arouse feelings of awe, wonder, and reverence. This aspect of drones is perhaps most evident in their ability to reinscribe natural sites and architectural forms with the sublime. Consider the growing popularity of drone photography featuring Niagara Falls or the Isle of Purbecks, for instance, natural sites routinely associated with sublime experiences. Likewise, drone footage of the Great Wall of China and the Golden Gate Bridge reinvest these monumental architectural achievements with the sublime. Significantly, many of the natural and architectural wonders David Nye (1994) discusses are perennial subjects for drone photography.

All of which is to suggest that drones are the latest in a series of technological innovations—from steamboats and railroads to jet aircraft—that not only make the sublime accessible to large publics, but also mediate sublime objects. Drones do so by producing dramatic aerial perspectives, commonly known as first-person view (FPV) footage, of both natural and architectural wonders. And, like earlier sublime technologies, the drone's visual association with Fourth of July fireworks displays—FPV clips of these celebrations frequently become viral sensations—rearticulates the relationship between technological innovation and national identity David Nye identifies in his analysis of the American technological sublime.

Of course, these are only brief illustrations of this book's central premise: that stories about technology play a decisive role in shaping how we develop, comprehend, and assimilate new technologies into our daily lives. In the following pages, we will examine how, why, and to what ends drones are discursively constructed within and through media discourse. We begin with an analysis of news discourse surrounding the introduction and rapid escalation of the US drone campaign.

Notes

1. See "The Drones," *Audi*, https://www.youtube.com/watch?v=OLSmkkw0v8I.
2. See "Amazon Prime Air," *Amazon*, https://www.youtube.com/watch?v=98BIu9dpwHU.
3. This evocative phrase is typically associated with Immanuel Kant's thinking on the sublime. For a concise discussion of the evolution of the concept see De Mul (2011).
4. A one-time student of Leo Marx, David Nye (1994) identifies several forms of the sublime: dynamic, mathematical, geometric, electric, and consumer. While Nye's attention to electrification centers on the transformation of factories and urban nightscapes, James Carey and John Quirk (1992) employ the phrase "electric sublime" to chart the evolution of electronic communication, from the telegraph to the computer.
5. In keeping with this study's discursive orientation, I will not offer a technological history of drones as such. For a concise discussion of the evolution of military drones, see Zaloga (2008).
6. Tesla received patent number US613809 for the "Method of and Apparatus for Controlling Mechanism of Moving Vessels or Vehicles" on November 8, 1898.
7. For a detailed discussion on this point, see Chapter 8: "What Inspires Them: Science Fiction's Impact on Science Reality" in Singer (2009).
8. For example, see "Video shows kangaroo knock drone out of sky," *UPI*, http://www.upi.com/Odd_News/2014/12/22/Video-shows-kangaroo-punching-drone-from-sky-destroying-the-device/2331419211009/.
9. See "They Watching Us," *DeWun Music*, https://www.youtube.com/watch?v=ZuH8n7Zsn_E.

References

Al Jazeera. (2015, March 4). French police continue hunt for Paris drones. *Al Jazeera English*. Retrieved from http://www.aljazeera.com/news/europe/2015/03/french-police-continue-hunt-mystery-paris-drones-150304104607724.html

Bogel, A. (2014). Birds attacking drones: Why these videos make us cheer. *Slate*. Retrieved from http://www.slate.com/articles/technology/future_tense/2014/10/birds_attacking_drones_why_these_videos_make_us_cheer.html

Carey, J. W., & Quirk, J. J. (1992). The mythos of the electronic revolution. In J. W. Carey *Communication as culture: Essays on media and society* (pp. 113–141). New York, NY: Routledge.

Cheney, M. (2001). *Tesla: Man out of time*. New York, NY: Touchstone.

Cohn, M. (Ed.). (2015). *Drones and targeted killing: Legal, moral, and geopolitical issues*. Northampton: Olive Branch Press.

Cole, C. (2013). Drone warfare and international law: Findings of UN reports on extrajudicial and arbitrary executions. *Global Research—Center for Research on Globalization*. Retrieved from http://www.globalresearch.ca/drone-warfare-findings-of-u-n-reports-on-extrajudicial-and-arbitrary-executions/5355601

Deckard, S. (2010). *Paradise discourse, imperialism, and globalization: Exploiting Eden*. New York, NY: Routledge.

Del Mul, J. (2011, July 17). The technological sublime. *Next Nature Network*. Retrieved from https://www.nextnature.net/2011/07/the-technological-sublime/

Del Prado, G. M. (2015, July 31). These weapons can find a target all by themselves—And researchers are terrified. *Business Insider*. Retrieved from http://www.businessinsider.com.au/which-artificially-intelligent-semi-autonomous-weapons-exist-2015-7

Diab, A. (2014). Drones perform the dull, dirty, or dangerous work. *Tech.Co*. Retrieved from http://tech.co/drones-dull-dirty-dangerous-2014-11

Dick, P. K. (2012). *Vulcan's hammer*. New York, NY: First Mariner.

Dunmire, P. (2007). "Emerging threats" and "coming dangers": Claiming the future for preventive war. In A. Hodges & C. Nilep (Eds.), *Discourse, war and terrorism* (pp. 19–43). Amsterdam: John Benjamins.

Dzieza, J. (2013, May 17). Why drones make us nervous. *The Daily Beast*. Retrieved from http://www.thedailybeast.com/articles/2013/05/17/why-drones-make-us-nervous.html

Foucault, M. (1984). The order of discourse. In M. Shapiro (Ed.), *Language and politics* (pp. 108–138). New York, NY: New York University Press.

Franklin, H. B. (2008). *War stars: The superweapon and the American imagination*. Amherst, MA: University of Massachusetts Press.

Future of Life. (2015). Open letter on autonomous weapons. *Future of Life Institute*. Retrieved from https://futureoflife.org/open-letter-autonomous-weapons

Gettinger, D., Holland Michel, A., Pasternack, A., Koebler, J., Musgrave, S., & Rankin, J. (2014). *The drone primer: A compendium of the key issues*. Center for the Study of the Drone. Retrieved from http://dronecenter.bard.edu/publication/the-drone-primer/

Gregory, D. (2011). From a view to a kill: Drones and late modern war. *Theory, Culture & Society*, *28*(7–8), 188–215.

Hackman, M., & Nicas, J. (2015, July 17). Drone delivers medicine to rural Virginia clinic. *Wall Street Journal*. Retrieved from https://www.wsj.com/articles/drone-delivers-medicine-to-rural-virginia-clinic-1437155114

Hall, M. (2007, May 29). Sci-fi writers join war on terror. *USA Today*. Retrieved from http://usatoday30.usatoday.com/tech/science/2007-05-29-deviant-thinkers-security_N.htm

Hornyak, T. (2015, May 12). Fujitsu drone uses wheels to roll along bridges, walls. *PC World*. Retrieved from http://www.pcworld.com/article/2921452/fujitsu-drone-uses-wheels-to-roll-along-bridges-walls.html

Kennedy, J. (1961, May 25). *Address to congress on urgent national needs*. Retrieved from http://www.jfklibrary.org/JFK/JFK-Legacy/NASA-Moon-Landing.aspx

Lombardo, T. (2006). *Contemporary futurist thought: Science fiction, future studies, and theories and visions of the future in the last century*. Bloomington, IN: AuthorHouse.

Marx, L. (1964). *The machine in the garden: Technology and the pastoral ideal in America*. New York, NY: Oxford University Press.

McNeal, G. S. (2012, April 4). A primer on domestic drones: Legal, policy and privacy implications. *Forbes*. Retrieved from http://www.forbes.com/sites/gregorymcneal/2012/04/10/a-primer-on-domestic-drones-and-privacy-implications/

Newcome, L. R. (2004). *Unmanned aviation: A brief history of unmanned aerial vehicles*. Reston, VA: American Institute of Aeronautics and Astronautics.

Newman, L. H. (2013, December 30). Five ways drones could do some good in 2014. *Slate*. Retrieved from http://www.slate.com/blogs/future_tense/2013/12/30/five_ways_drones_unmanned_aerial_vehicles_could_work_with_nonprofits_in.html

Nye, D. E. (1994). *American technological sublime*. Cambridge, MA: MIT Press.

RT News. (2015, July 17). Gunslinger drone, your moderately-priced assassin. *RT.com*. Retrieved from https://on.rt.com/6n9t

Scollon, S., & Scollon, R. (2000). *The construction of agency and action in anticipatory discourse: Positioning ourselves against neo-liberalism*. Paper presented at the Third Conference for Sociocultural Research UNICAMP, Sao Paulo, Brazil.

Shadbolt, P. (2015, January 14). Robo-wings: Military drones mimic hawks and insects. *CNN.com*. Retrieved from http://www.cnn.com/2015/01/14/tech/mci-drone-robohawk-robofly

Shah, P. Z. (2012, February 27). My drone war. *Foreign Policy*. Retrieved from http://foreignpolicy.com/2012/02/27/my-drone-war/

Singer, P. W. (2009). *Wired for war: The robotics revolution and conflict in the 21st century*. New York, NY: Penguin Press.

Sofke, E. (2015). What might a killerbot arms race look like? *Popular Science*. Retrieved from http://www.popsci.com/what-would-killerbot-arms-race-look

Stahl, R. (2013). What the drone saw: The cultural optics of the unmanned war. *Australian Journal of International Affairs*, *67*(5), 658–674.

Streeter, T. (1997). Blue skies and strange bedfellows: The discourse of cable television. In L. Spigel & M. Curtin (Eds.), *The revolution wasn't televised: Sixties television and social conflict* (pp. 221–242). London: Routledge.

Tesla, N. (2010). *The problem of increasing human energy: With special reference to harnessing the sun's energy*. New York, NY: Cosimo Classics.

Thompson, L. (2015, April 10). Air force's secret "gorgon stare" program leaves terrorist nowhere to hide. *Forbes*. Retrieved from http://www.forbes.com/sites/lorenthompson/2015/04/10/air-forces-secret-gorgon-stare-program-leaves-terrorists-nowhere-to-hide/

Turi, J. (2014, January 19). Tesla's toy boat: A drone before its time. *Engadget*. Retrieved from http://www.engadget.com/tag/timemachine/www.engadget.com/2014/01/19/nikola-teslas-remote-control-boat/

Watson, I. (1985). *Slow birds and other stories*. London: Victor Gollancz.

Weaver, M. (2014, May 11). RSPB uses drone to keep watch on Britain's vulnerable birds. *The Guardian*. Retrieved from https://www.theguardian.com/environment/2014/may/11/rspb-drone-britain-vulnerable-birds?CMP=twt_gu

Weber, A. (2010). Nikola Tesla: Father of unmanned vehicle technology. *Assembly Magazine*. Retrieved from http://www.assemblymag.com/articles/87689-nikola-tesla-father-of-unmanned-vehicle-technology

Zaloga, S. J. (2008). *Unmanned aerial vehicles: Robotic air warfare 1917–2007*. Oxford: Osprey Publishing.

· 2 ·

A NEW KIND OF WAR

The best way to keep the peace is to redefine war on our terms.
—George W. Bush, Campaign Speech at The Citadel, September 23, 1999

In March 2009, the Department of Defense issued a memo to Pentagon staff calling for an end to the war on terror. To be clear, the directive did not call for the end of hostilities. Rather, the recently installed Obama administration sought to rebrand America's counterterrorism efforts. The memo explained, "this administration prefers to avoid using the term Long War or Global War on Terror … please pass this on to your speechwriters." Favoring technocratic language over the bellicose rhetoric of his predecessor, George W. Bush, the new president's Defense Department urged Pentagon officials to instead use the phrase "Overseas Contingency Operations" (Wilson & Kamen, 2009). This was not the first time the war on terror received a discursive makeover. During Bush's second term, Secretary of Defense Donald Rumsfeld floated the label "global struggle against violent extremism." Like Rumsfeld's earlier linguistic intervention, the Obama administration's name change failed to gain much traction with journalists.

In subsequent years, Obama continued to distance himself from Bush's terminology, but to no avail. As a matter of course, news workers and analysts use the phrase "war on terror" to discuss US military interventions from Central Asia and the Middle East to the Arabian Peninsula and the Horn of Africa. Despite the failure of Obama's discursive strategy, there is a cruel irony to this rebranding effort; between 2009 and 2015 Obama ordered upwards of 500 drone strikes, ten times the number conducted by the Bush administration (Press TV, 2015). Small wonder, then, that reporters and critics from across the political spectrum frequently described drones as Barack Obama's "weapon of choice."

> The CIA currently wages a 24/7 Predator campaign against the Pakistani Taliban and Al Qaeda. In Pakistan, drone attacks are Obama's weapon of choice. (*Newsweek*, June 7, 2010)

> Since he came to the presidency in 2008, Obama and his officials have increasingly adopted the controversial drone program as their weapon of choice against leaders of militant groups. (*Vancouver Sun*, September 23, 2011)

> Despite President Barack Obama's recent call to reduce the United States' reliance on drones, they will likely remain his administration's weapon of choice. (*Foreign Affairs*, June 11, 2013)

> From his first days as commander in chief, the drone has been President Barack Obama's weapon of choice, used by the military and the CIA to hunt down and kill the people his administration has deemed—through secret processes and without trial—worthy of execution. (*The Intercept*, October 15, 2015)

Having dramatically escalated his predecessor's use of drones for targeted assassination, these killing machines are, to borrow a headline from *USA Today*, "the face of the war on terror" (Michaels, 2015). Put differently, drones function as a *synecdoche* for the war on terror. And yet, it is far from clear how drones moved from relative obscurity to assume such prominence in military affairs, let alone the popular imagination.

In an effort to plot the rise of the drones, in US counterterrorism strategy as well as public consciousness, this chapter examines media discourse on the secretive drone program during the Bush administration. Chapter 3 continues this line of inquiry under the Obama presidency. Rather than attempt a comprehensive analysis of this material, I employ the concept of *critical discourse moments* (Chilton, 1987; Gamson, 1992) to analyze news coverage of key episodes in America's drone wars. Doing so, I aim to highlight "various stages of the public life of such issues, from the conditions of emergence in the public arena to their constitution into political problems" (Carvalho, 2010,

p. 13). Throughout, I chart the evolution of the "public life" of drone warfare under Presidents Bush and Obama. Adopting a "discourse-historical" approach (Wodak, 1999), these chapters examine how the meanings of drones and targeted killing emerged and developed over the course of the so-called war on terror. We begin with the first deployments of weaponized drones during the Bush years.

A Period of Consequences

Early in the 1999 Republican presidential primaries, George W. Bush emerged as the party's frontrunner. The Texas governor proved to be a formidable fundraiser whose social policy and domestic agenda resonated with conservative voters. Despite Bush's name recognition and surging popularity, the candidate's lack of foreign policy credentials presented a major obstacle to his securing the nomination. GOP rivals and Democratic opponents alike questioned Bush's aptitude and experience in international relations and military affairs. In a move designed to counter those criticisms, Bush gave his first major defense policy address at The Citadel, a historic military college in Charleston, South Carolina. The setting, along with the national security team Bush had assembled, was meant to dispel fears that he was unprepared to become America's next commander-in-chief. More important for our purposes here, Bush's speech, and the attendant news coverage, constitutes the first in a series of *critical discourse moments* that would usher in the era of drone warfare.

Gamson describes critical discourse moments as key episodes "that make discourse on an issue especially visible. They stimulate commentary in various public forums by sponsors of different frames, journalists, and other observers" (1992, pp. 25–26). As we shall see, Bush's address, "A Period of Consequences," served as a reference point for subsequent discussion of the character and conduct of the war on terror. More immediately, the speech brought ongoing developments in high-tech weaponry to light. General Atomics' RQ-1 Predator aircraft, perhaps the most recognizable unmanned vehicle in America's growing drone arsenal, had been operational for reconnaissance and surveillance missions since the 1990s, albeit with little fanfare. And in the waning days of his presidency, Bill Clinton's counterterrorism chief, Richard Clarke, quietly but persistently urged the Central Intelligence Agency (CIA) to arm these surveillance planes in the hunt for Osama bin Laden, who was wanted at the time for al Qaeda's role in the 1998 US embassy bombings in Tanzania and Kenya (Whittle, 2014). Thus, while unmanned planes were proving to be valuable tactical assets, these remotely piloted vehicles were relatively unknown prior to the US invasion of Afghanistan, the first battleground in America's war on terror.

Speaking before an enthusiastic crowd of cadets, George W. Bush outlined his vision for American defense policy in the new century. Throughout, Bush identified several challenges confronting the US military, principally, "a world of terror and missiles and madmen" (Bush, 1999). Vowing to defend the homeland, Bush promised that as president he would "begin creating the military of the next century." Like generations of leaders before him, Bush sought a technological fix to America's most daunting problems: as he saw it, the twin perils of rogue states and transnational terror networks. In a passage that anticipates killing by remote control on a global scale, Bush told the cadets, "In the air, we must be able to strike from across the world with pinpoint accuracy—with long-range aircraft and perhaps with unmanned systems." Given the purchase of the technological sublime on the journalistic imaginary, press accounts highlighted Bush's call for new weapon systems that would "skip a generation."

> **Bush would emphasize next-generation weaponry, missile defenses**
> (*AP*, September 23, 1999)
>
> **Bush vows to upgrade futuristic arms research**
> (*St. Louis Post-Dispatch*, September 24, 1999)
>
> **BUSH VOWS MORE DEFENSE MONEY THE GOP FRONT-RUNNER LAID OUT A GET-TOUGH FOREIGN POLICY. HE CALLED FOR $20 BILLION OVER 5 YEARS FOR HIGH-TECH ARMS**
> (*Philadelphia Inquirer*, September 24, 1999)

Following the speech, Bush refused to take questions, referring journalists instead to Condoleezza Rice, a rising star in national security circles, for details on his defense policy. Moreover, the Bush campaign made it clear that the candidate's call for military transformation had the blessing of former national security staff—Dick Cheney, Paul Wolfowitz, Richard Armitage and Donald Rumsfeld—like Ms. Rice, all of whom served Bush senior; and all of whom would play decisive roles in the war on terror. The frequency with which these figures appear in news accounts of the Citadel speech reveals the Bush campaign's discursive practice and the strategy behind it: name-dropping his father's security advisors in an effort to bolster the candidate's defense credentials and legitimate his plans for "a revolution in the technology of war" (Bush, 1999). Bush's discursive strategy proved effective, especially among neoconservatives. Consider this fawning exchange between Fox News analysts Fred Barnes and Morton Kondracke.

> BARNES: You know, this was really a great speech, Mort. I mean, it could have been written by his hardline advisers, Richard Perl and …
> KONDRACKE: Paul Wolfowitz.
> BARNES: Paul Wolfowitz, exactly.
> KONDRACKE: Cheney.
> BARNES: And they may—and Dick Cheney. Yes, they—and they all, I think, participated in writing it. The emphasis was on creating, building a new generation of high-tech weapons that will leave something that potential enemies wouldn't be able to match for years or even a decade or more. (Fox News, 1999)

Bush's "Period of Consequences" speech, designed to allay fears of the candidate's inexperience in defense and national security, achieved the desired effect. Curiously, from that point forward, Bush rarely discussed terrorism or his ambitious plans to transform the US military—on the campaign trail at any rate. Before long however, the attacks of September 11, 2001, and the Bush administration's jingoistic response to those crimes, would shed new light on the Citadel address, effectively recasting his remarks as a form of anticipatory discourse: one that foresaw the terrorist threat and compelled the United States to "redefine war on our terms" (Bush, 1999).

The rapid development and subsequent deployment of weaponized drones in the skies above Afghanistan epitomized Bush's call to redefine America's wars. During the first months of Operation Enduring Freedom—the Pentagon's official name for the Afghan War[1]—news accounts of drone strikes appeared with increasing frequency. Fresh from a joint operation between an armed drone and two Navy jet fighters that killed Mohammed Atef, bin Laden's heir apparent and alleged mastermind of the 9/11 attack, America's fleet of Predators quickly became an object of media fascination. A story profiling the Predator in the New York *Times*, pegged to Atef's death, is indicative of this growing trend.

Headlined "Ugly Duckling Turns Out to Be Formidable in the Air," the *Times* describes the unmanned vehicle as an "ungainly, propeller-driven craft that flies as slowly as 80 miles per hour and is operated by remote control," adding, "military experts say the ability to fly these armed, unmanned drones is a significant advance for search-and-destroy missions on the battlefield" (Miller & Schmitt, 2001). The report goes on to provide a brief history of the

Predator: its origins in an Israeli design; its first combat deployment in Bosnia; and the plane's reconfiguration from surveillance platform to lethal weapon under the auspices of defense contractor General Atomics.

Midway through the story, the narrative focus shifts from the aircraft's technical provenance to administrative infighting between the Pentagon and the CIA over the aircraft's command and control. The *Times* recounts an early episode in the Afghan War in which suspected al Qaeda and Taliban leaders were surveilled by a Predator, and the complicated and confused decision making process that led to one of the first (unsuccessful) drone strikes in US military history.

> The call went out from the Central Command's sophisticated operations center near Riyadh, Saudi Arabia, to General Franks's command's headquarters in Tampa, Fla., and to Defense Secretary Donald H. Rumsfeld in Washington. Mr. Rumsfeld gave the order to attack, and called President Bush to alert him of the developments, a military official said. By the time approval came back, part of the convoy had driven off. (Miller & Schmitt, 2001)

As related by the *Times*, the episode inspires a sense of awe at the geographic scope and techno-bureaucratic scale of America's nascent targeted killing program. Equally important, the *Times* offers readers a reassuring sense of ideological closure by noting that tensions between the Pentagon and the CIA have since been resolved under a "new agreement" between the military and the spy agency. Returning to the vehicle's technical capabilities, the story concludes with the Pentagon's acknowledgement of the aircraft's shortcomings: the Predator's communication links are susceptible to interference in adverse weather conditions. More fundamental questions over the accuracy of the intelligence used to track individuals, let alone the precision of the Predator's targeting and weapons systems, are neither raised nor addressed.

Notwithstanding increased media attention to the use of armed drones in America's war on terror, it was President George W. Bush's return to The Citadel, on December 11, 2001, the third-month anniversary of the 9/11 attacks, that served as the Predator's coming out party. Bush's second address at The Citadel, wherein he singled out the aircraft for its revolutionary potential, generated considerable news coverage and increased public awareness of these unmanned vehicles. As such, Bush's reprise at The Citadel constitutes a second critical discourse moment in the public life of drone warfare.

Speaking at a news conference prior to Bush's return to the prestigious military college, Press Secretary Ari Fleischer told reporters: "The President's speech tomorrow at The Citadel will be an interesting follow-up to a speech that he gave there in the fall of 1999. The speech also, of course, coincides with the three-month anniversary of the attack on our country" (Press Briefing, 2001). Fleischer went on to explain the White House was distributing the text of Bush's previous address and told reporters that that speech contained "the seeds" of the President's approach to combatting terrorism. As an exercise in strategic communication, Fleischer's briefing not only supplied journalists with the administration's preferred framing for the President's remarks, but also primed reporters to contextualize the speech in light of the Bush administration's day-long commemoration of the 9/11 attacks. An obliging press corps constructed narratives that featured the President's Citadel speech in the context of memorials in New York City and Washington, DC earlier in the day (e.g., Hutcheson, 2001a).

A handful of news workers observed the media orchestration behind Bush's return to The Citadel. For example, in the lead to his report, Mike Allen (2001) noted: "The administration took the unusual step yesterday of e-mailing reporters the text of a national security speech Bush had delivered in 1999, when he was Texas governor and beginning his fight for the Republican presidential nomination." Speaking on ABC's *Good Morning America*, George Stephanopoulos told viewers:

> Two years ago, September 1999, Bush gave his first speech when he was running about terrorism. And his first [speech] had the first explanation of the Bush doctrine, that if you harbor a terrorist, you're going to be attacked. The Bush White House is putting this out, saying it shows that Bush was very prescient, but that was only one speech given in the campaign. (ABC News, 2001)

Meanwhile, Los Angeles *Times* reporter Ron Brownstein told CNN, the President "was quite prescient also in suggesting that the contests and the challenges of the future would be very different than the Gulf War" (CNN Live Today, 2001). For the most part, however, reporters either downplayed or ignored the administration's "unusual" discursive practice of distributing the text of the previous address: a calculated effort designed to emphasize the foresight of Bush's "A Period of Consequences" speech.

The majority of news reports emphasized the prominent role new technology was playing in the three-month-old war on terror. Here, the Bush administration's release of the previous speech proved especially effective in shaping coverage of the President's return to The Citadel. In addition to demonstrating the intertextual character of political and news discourse, these stories routinely employ Bush's rhetorical trope of military transformation through technological revolution.

> "I said here at the Citadel … America was entering a period of consequences that would be defined by the threat of terrorism, and that we faced the challenge of military transformation," Bush told some 2,000 cadets whose white gloves muffled their applause. (Sobieraj, 2001)

> "This revolution in our military is only beginning, and it promises to change the face of battle," the president told about 2,000 cadets at The Citadel, a South Carolina military college. (Hutcheson, 2001b)

> "When all of our military can track moving targets with surveillance from air and space, warfare will be truly revolutionized," Bush said. "The need for military transformation was clear before the conflict in Afghanistan and before Sept. 11." (Heath, 2001)

Describing the tactical advantages afforded by new weapon systems—a combination of real-time intelligence and precision ordnance—Bush mentioned the Predator by name only once. Nevertheless, journalists seized on the President's claim that unmanned systems and related innovations would not only transform the military but also change the nature of armed conflict.

> [Bush] spoke about how innovative doctrine and high technology are rewriting the rules of war. He talked about the transformation of the US military and how Afghanistan is a proving ground for these new, innovative techniques. "War is teaching us more about the future of the military," he said, "than decades of blue ribbon panels." He spoke about the usefulness of the Predator, precision-guided munitions, other new kinds of technologies, ability to locate and target moving targets. (Conan, 2001)

> Citing the experience of the US in Afghanistan in rewriting the rules of war, combining techniques from land and air in a way they had never before achieved, he called on politicians and the military to abandon pet projects and old ways in the cause of fundamental reform. (Smyth, 2001)

> In his speech, Mr. Bush pointed to the Afghanistan operation as a "proving ground" for using high-tech weapons to win an unconventional war. He called upon the military to revolutionize its forces with unmanned aircraft and space surveillance. (Anon, 2001)

Writing a year in review piece, veteran journalist Robert G. Kaiser neatly summarized how the combination of new adversaries, new weapons, and new command and control structures were indeed redefining America's wars.

> The conflict in Afghanistan has included several "firsts" for an American war: This was the first conflict in which unmanned Predator drones firing Hellfire missiles have been used; the first to feature extensive use of a CIA paramilitary force as the first armed American unit on the ground; the first deployment by US forces to Central Asia; and apparently the first foreign war in which real-time approval from headquarters in the United States was required before weapons could be fired at certain targets that were in plain sight—suspected Taliban and al Qaeda leaders. (Kaiser, 2001)

While journalists' fascination with these new weapon systems is clear, questions regarding the legality, let alone the efficacy of this new kind of war, were conspicuously absent from news discourse at this time. Following several high-profile drone strikes in the coming year, those issues would eventually surface, raising a series of vexing legal, moral, and strategic questions that continue to inform debates over the use of unmanned systems in declared and undeclared war zones (Chamayou, 2015). These episodes constitute a series of critical discourse moments in the increasingly controversial public life of the drone.

Is it Bin Laden?

In February 2002, news stories about a drone strike in the remote Zhawar Kili region of eastern Afghanistan, close to the Pakistan border, focused on the possibility that Osama bin Laden, America's most wanted fugitive, had been killed. According to published reports, the attack commenced soon after an armed Predator spotted a "tall man" surrounded by a group of deferential followers in an area once controlled by al-Qaeda.[2] The story made international headlines, albeit with qualifications and descriptions that reveal the discursive practices and ideological positions of the respective news organizations.

> **Afghan official says seven people killed in US missile strike – possibly including an al-Qaida leader**
>
> (*AP*, February 7, 2002)
>
> **US officials refuse to comment on reports about death of top al-Qaeda**
>
> (*AFP*, February 7, 2002)
>
> **Senior Al-Qaeda official thought killed in attack**
>
> (*Deutsche Presse-Agentur*, February 7, 2002)
>
> **CIA missile kills 'al Qaeda leaders'**
>
> (*The Herald* (Scotland), February 7, 2002)
>
> **AL QAEDA BIGS NAILED BY MISSILE**
>
> (*New York Post*, February 7, 2002)

Both the Associated Press (AP) and Agence France Presse (AFP) headlines refer to official sources, thereby attributing the story to unnamed American or Afghan authorities. By contrast, the Deutsche Presse-Agentur offers the qualification "thought killed" without directly attributing this assessment to anyone in particular, whereas the unattributed quotes in *The Herald* headline make implicit reference to a knowledgeable, but unidentified source. Finally, the tabloid New York *Post* offers no qualifications in a headline that scarcely conceals glee over the air strike.

Subsequent reports, fueled by speculation of bin Laden's death—the headline in Australia's *Courier Mail* "Is it Bin Laden?" is indicative of such stories—soon cast doubt on the veracity of such claims and raised questions about the intelligence used to launch the assault.

> Allegations of deadly errors in two recent high profile attacks aimed at what were believed to be al-Qaeda targets in Afghanistan raised more questions Monday about the reliability of US intelligence.
>
> Villagers said three people killed in a CIA-launched missile strike February 4 in eastern Afghanistan were "hungry" peasants scavenging for scrap metal in a former al-Qaeda stronghold, and not al-Qaeda, Afghanistan's deputy border affairs minister Mirza Ali said. (Mannion, 2002)

> Afghan villagers told three American newspapers that the victims were peasants scavenging for scrap metal on a remote, snowcapped mountainside in Zhawar Kili when a US Hellfire missile, launched by an unmanned Predator drone, shrieked down on them a week ago. (MacKenzie, 2002)

> Three men killed in a rocket attack by US forces in Afghanistan last week were not members of the al-Qaeda terrorist network and Osama bin Laden was not among them, the Washington *Post* reported Monday. (Deutsche Presse-Agentur, 2002)

Notwithstanding the dramatic flourishes of MacKenzie's account in the Ottawa *Citizen*, the sourcing for these stories further underscores the discursive practice of intertextuality in news discourse. That is, while the AFP story attributes the description of the victims to an Afghan official, others draw on reports from "American newspapers," with specific reference to the Washington *Post* in the third example above. While this attribution lends credibility to these accounts, a closer look at US reporting reveals a far more complicated picture: one that deflects attention away from faulty intelligence and justifies the drone strike, regardless of the victims' true identities.

Headlined "Casualties of US Miscalculations; Afghan victims of CIA missile strike described as peasants, not Al Qaeda" the Washington *Post* frames the attack as a regrettable, but unavoidable error, effectively diminishing American culpability for the loss of innocent lives (Struck, 2002). Offering neither apologies nor excuses, a New York *Times* story leads with the Pentagon's justification for the deadly assault, including the military's rebuttal to claims of intelligence failures.

> Senior military officials today defended a missile attack by a Central Intelligence Agency drone over eastern Afghanistan last week, calling the strike "appropriate" and sharply questioning assertions by Afghans that the missile killed as many as three peasants who were merely scrounging for scrap metal. (Dao, 2002)

Similar storylines appeared in domestic and international news outlets. The following paragraph, taken from a foreign wire service, is indicative of Pentagon spin uncritically picked up and circulated by Western news organizations.

> Troops who searched the site also recovered documents, including airline schedules and English language credit card applications, as well as what appeared to be communications equipment and small arms and ammunition, said Rear Admiral John Stufflebeem, deputy operations director of the Joint Staff. (Agence France Presse, February 2002)

As the Pentagon spokesperson on the incident, Stufflebeem not only disputed the dead's innocence, but also claimed "something untoward" (Stone, 2002) was underway, going so far as to suggest that the controversial drone strike foiled an imminent terror plot. Thus, Pentagon spin successfully deflected attention away from intelligence failures. Instead, journalists turned their attention to dramatic efforts to recover evidence from the site of the attack and the collection of DNA samples to determine whether or not bin Laden was killed.

> **US Seeks Osama DNA from family**
>
> (*New York Post*, February 28, 2002)
>
> **US asks Bin Laden family for DNA**
>
> (*Calgary Herald*, February 28, 2002)
>
> **Has bin Laden died in Hellfire? DNA checks on missile victim in the Afghan peaks**
>
> (*Western Daily Press*, February 28, 2002)

While it is uncertain whether the Pentagon engaged in a disinformation campaign to deflect criticism of the drone program, it is clear that many news outlets uncritically accepted the military's account, and in some instances, unequivocally supported the fatal assault. Later that year, a drone strike in Yemen, far removed from the declared battlefields of Afghanistan, would raise fresh questions about the US drone program, thereby constituting another critical discourse moment—one that would spell out for the first time the terms of the debate surrounding the use of drones in America's expanding war on terror.

The Dawn of Robotic War

Qaed Salim Sinan al-Harethi, the target of a lethal November 4, 2002 drone strike in Yemen, was a senior al-Qaeda operative believed responsible for the deadly attack on the USS Cole two years earlier. Five other men, including an American citizen, whose names were rarely mentioned in initial news reports, died along with al-Harethi as they traveled by jeep along a remote road in Marib province, 100 miles northeast of the Yemeni capital, Sana'a. Subsequent reports identified the dead American as Kamal Derwish, a dual Yemeni-US citizen thought to be part of a "homegrown" terror cell based outside of Buffalo, New York. According to Bush administration officials, Derwish was not targeted in the strike. Nonetheless, the White House asserted its legal authority to kill an American working with al-Qaeda; a claim that went largely unchallenged in news discourse—a stark contrast to the controversy surrounding the 2011 killing of Anwar al-Awlaki discussed at length in the following chapter.

News accounts of the drone strike that killed al-Harethi followed a pattern that has since become routine in reporting the drone wars. For instance, the majority of news reports on al-Harethi's death duly noted the CIA's refusal to comment, followed in short order by an official statement from an unnamed source. An Associated Press International report is indicative of this reliance on such sources: "A US official in Washington confirmed Monday that US forces carried out the strike, believed to have been conducted by a CIA aircraft, possibly a missile-carrying Predator drone. The official spoke on condition of anonymity" (Al-Haj, 2002). Conversely, print and broadcast news stories invariably included statements from senior Bush administration officials. For example, Defense Secretary Donald Rumsfeld, who neither confirmed nor denied US involvement in the attack, nevertheless offered his opinion on the outcome of the operation. Speaking at a press conference, Rumsfeld told reporters "it would be a very good thing if he [al-Harethi] were out of business" (Martin, 2002). Meanwhile, Deputy Defense Secretary Paul Wolfowitz described the drone strike as "a very successful tactical operation" (Kagan & Ensor, 2002). All of which illustrates that reliance upon official sources—anonymous or otherwise—is a long-standing discursive practice for government figures and news workers alike: a practice which, as we shall see in Chapter 6, shapes coverage of the drone wars in fundamental and decisive ways.

Despite the routinized nature of these reports, however, there are some distinguishing characteristics in accounts of the Yemen drone strike that merit further scrutiny. For example, elaborating on the trope of technological revolution discussed above, headlines describing al-Harethi's death are replete with references to "robots" and "robotic war." These descriptions echo and amplify earlier assertions, made by government officials and journalists alike, that drones are redefining modern warfare.

> **Robotic warfare leaves terrorists no hiding place**
> (*The Times*, November 6, 2002)
>
> **US Hails Ambush by Robot**
> (*The Sun*, November 6, 2002)
>
> **Quiet assassin—Robot plane's deadly strike heralds new era of warfare**
> (*Sunday Mail*, November 10, 2002)
>
> **The Year in Ideas: Robotic Warfare**
> (*New York Times*, December 15, 2002)

In another instance of intertextuality in news discourse, a quote from Clifford Beal, editor of *Jane's Defence Weekly*, may account for this characterization of drone technology. Dozens of reports included reference to Beal's assessment of the drone strike that killed al-Harethi: "To have a drone that engages and kills people—that is quite a threshold to cross. … This is the beginning of robotic warfare" (Lathem & Sujo, 2002). This characterization has since become a prominent trope in discussions of drone warfare, one that connects and resonates with the discourse of the technological sublime discussed in Chapter 1.

Beal's further assertion regarding parallels between the US drone strike and the controversial Israeli tactic of targeted killing gained far less traction. Nevertheless, several international reports went on to quote Beal. "It doesn't seem they [al-Harethi's group] were given the opportunity to surrender. They were taken out Israeli-style" (Buncombe & Whitaker, 2002). Veteran journalist Robert Fisk (2002) observed that it was in the context of al-Harethi's death that US officials first began to employ "Israel's own mendacious description of such murders as a 'targeted killing,'" to discuss America's drone assassination program. As a result, comparisons between Israeli and US counterterrorism strategies emerged in follow up reporting on al-Harethi's death. The resultant debate raised questions over possible violations of human rights and international law in America's drone campaign.

With greater frequency, journalists and critics called attention to US denunciations of the Israeli practice of targeted killing, leading to charges of American hypocrisy. For instance, a report published by the Inter Press Service noted: "The incident, which sparked outrage in Arab countries, immediately drew questions about parallels with Israel's policy of 'targeted killings' of suspected Palestinian terrorists, a policy condemned even by the Bush administration" (Lobe, 2002). Investigative journalists Tony Geraghty and David Leigh went on to suggest,

> Israeli hardliners had the pleasure this week of seeing their controversial tactic of "targeted killing" of their enemies vindicated by being imitated. … The US has borrowed not just their policy, but their techniques too. It was Israel that pioneered the use of the Hellfire missile for summary executions such as the US carried out last month in Yemen. (Geraghty & Leigh, 2002)

The international press was not alone in observing a correspondence between Israel's targeted killing program and the US drone campaign; domestic news organizations began asking tough questions as well. Consider this passage from *Newsweek* magazine: "Reporters were starting to ask: was the United States getting into the murky business of assassination? How was this kind of hit job different from the Israeli policy of 'targeted killings' of Palestinian militants, denounced by Washington?" (Thomas & Hosenball, 2002). In response, Bush administration officials attempted to draw distinctions between America's drone program and the Israeli tactic of targeted killing: a practice US officials continued to denounce, even in the aftermath of al-Harethi's assassination.

Addressing charges of double standards, State Department spokesman David Boucher unequivocally defended the US position.

> Asked how the action in Yemen squares with past US opposition to Israeli "targeted killing" of Palestinians, Boucher said, "If you look back at what we have said about targeted killings in the Israeli-Palestinian context, you will find that the reasons we have given do not necessarily apply in other circumstances." (Pincus, 2002)

Despite Boucher's suggestion, *Post* reporter, Walter Pincus, does not "look back" at previous administration statements regarding Israel's practice of targeted killing. And, like Boucher, Pincus fails to elaborate on what, precisely, the "reasons" and "circumstances" might be for America's apparent violation of international law. In an effort to refute such accusations, succeeding official statements highlighted the close cooperation between US and Yemeni forces in conducting a precision drone strike. Unsurprisingly, the majority of follow up reports uncritically repeated official denials of any contradictions in American policy. Likewise, opinion pieces echoed the administration's talking points. For instance, a Denver *Post* editorial described the CIA drone strike as "a good shot," adding that in light of Yemen's cooperation in the lethal assault, the drone strike did not violate Yemeni sovereignty. As media critic Norman Solomon wryly observes of such journalistic practices, "Under the unwritten laws of American media coverage, such denials tend to end the matter" (2005, p. 59).

For instance, former CIA counsel, Jeffrey Smith, offered qualified support of the US drone campaign. During an interview with NPR, Smith acknowledged historical precedents of state-sanctioned assassinations by the Israelis in the Occupied Territories and the British in Northern Ireland. Nonetheless, questions of strategic efficacy and political expediency outweighed legal, moral, and human rights concerns, as in this exchange between Smith and television reporter David Shuster.

> SHUSTER: Ethics and double standards aside, some analysts argue that targeted killings don't always work.
>
> SMITH: There is always a risk when you use lethal force that it's going to enflame passions, and that's why I believe that any time the United States wants to make a strike like this, it has to do so very carefully, weighing all of the factors and the decision has to be made at a sufficiently high level so that all of these factors can be weighed. (Matthews, 2002)

Thus, domestic news reports and critical commentary downplayed the legal and ethical implications of targeted killing, thereby lending tacit approval to the Bush administration's expansion of the US drone campaign. Conversely, dissenting perspectives, including those of the human rights organization Amnesty International and Swedish Foreign Minister, Anna Lindh, who condemned the strike as a "summary execution that violates human rights," were only occasionally referenced in news accounts. More often than not, such criticism was buried at the end of these stories, frequently without benefit of a direct quote (e.g., Pincus).

Weeks later, the New York *Times* reported al-Harethi was included on a recently disclosed list of "high-value targets" compiled by the Bush administration: a list that authorizes the CIA to kill terror suspects in the event that capture proves impractical and civilian casualties can be limited. According to the *Times*, "The president is not legally required to approve each name added to the list, nor is the CIA required to obtain presidential approval for specific attacks, although officials said Mr. Bush had been kept well informed about the agency's operations" (Risen & Johnston, 2002). Here the *Times* uncritically repeats official statements claiming the targeted killing program is a tactic of last resort, one that is subject to thorough and judicious oversight. What's more, the *Times* glosses over contradictions in American policy with the inclusion of a statement from yet another unnamed source, who declares that the two dozen terror suspects on the list are "the worst of the worst"—an assessment, dutifully repeated in subsequent reporting, which not only defies the rule of law but also legitimates the increasingly common practice of extra-judicial killing.

Unlike the controversial drone strike in Yemen that raised questions over human rights and international law, this latest revelation failed to attract much media attention, let alone stir substantive debate. By late 2002 and early 2003, news organizations focused their energies and resources on the run up to US-led invasion of Iraq. Significantly, drones were part of that story too, albeit for different reasons and to different ends. In making its case for invading Iraq, the Bush administration claimed Saddam Hussein planned to use unmanned aircraft to deliver weapons of mass destruction (WMD) to targets on the US mainland. Like intelligence reports on Iraq's WMD program, these claims have since been debunked. Nevertheless, the ruse appears to have convinced several Senate Democrats to support Bush's war plans (Friedersdorf, 2012).

Meanwhile, the White House continued its drone campaign in Afghanistan; and, beginning in 2004, commenced CIA-directed drone strikes in Pakistan, yet another undeclared battleground in an ever-expanding war on terror. Nevertheless, during the invasion and subsequent occupation of Iraq, coverage of the US drone program took a backseat to more conventional forms of warfare. It would remain for the historic election of a new president—who campaigned as an anti-war candidate and famously called for greater government transparency and respect for the rule of law—for drones to return to the spotlight as Barack Obama's "weapon of choice."

Notes

1. Initially, war planners dubbed the Afghan invasion Operation Infinite Justice, but changed the name after learning that Muslims found the label offensive. See Solomon (2005, p. 134).
2. News organizations use different spellings for the terrorist group. Quotes taken from news accounts retain each news outlet's preferred spelling.

References

ABC News. (2001, December 11). How the September 11 terrorist attacks changed President George W. Bush. *Good Morning America. Lexis-Nexis Universe: General News Topics.*

Agence France Presse. (2002, February 11). US has no information to conclude missile strike was a mistake: Pentagon. *Agence France Presse-English. Lexis-Nexis Universe: General News Topics.*

Al-Haj, A. (2002, November 4). Six al-Qaida suspects killed in car explosion in Yemen, officials say. *Associated Press International. Lexis-Nexis Universe: General News Topics.*

Allen, M. (2001, December 11). Bush talk to mark attack on 11th. *The Washington Post.* Retrieved from https://www.washingtonpost.com/archive/politics/2001/12/11/bush-talk-to-mark-attacks-on-11th/987f106c-9dae-4f63-8e0b-19acc19a6c12/?utm_term=.712987448b97

Anon. (2001, December 12). Pax American? *St. Louis Post-Dispatch*, Editorial, p. C16. *Lexis-Nexis Universe: General News Topics.*

Buncombe, A., & Whitaker, R. (2002, November 7). Deadly revenge from the office. *New Zealand Herald*, World. *Lexis-Nexis Universe: General News Topics.*

Bush, G. W. (1999, September 23). *A period of consequences.* Speech delivered at the Citadel, Charleston, SC. Retrieved from http://www3.citadel.edu/pao/addresses/pres_bush.html

Carvalho, A. (2010). Media(ted) discourse and society: Rethinking the framework of critical discourse analysis. In J. Richardson (Ed.), *Language and journalism* (pp. 10–26). New York, NY: Routledge.

Chamayou, G. (2015). *A theory of the drone.* New York, NY: The New Press.

Chilton, P. (1987). Metaphor, euphemism, and the militarization of language. *Current Research on Peace and Violence, 10*, 7–19.

CNN Live Today. (2001, December 11). President Bush's military vision then and now. *CNN.com.* Retrieved from http://www.cnn.com/TRANSCRIPTS/0112/11/lt.21.html

Conan, N. (2001, December 11). President Bush's speech at The Citadel and reaction to the speech. *NPR*, Talk of the Nation. *Lexis-Nexis Universe: General News Topics.*

Dao, J. (2002, February 11). US defends missile strike, saying attack was justified. *New York Times*, Section A, p. 10. *Lexis-Nexis Universe: General News Topics.*

Deutsche Presse-Agentur. (2002, February 11). Man killed in Afghan attack was not Bin Laden, reports paper. *Lexis-Nexis Universe: General News Topics.*

Fisk, R. (2002, November 8). Bush fights for another clean shot in his war. *The Independent*, Comment, p. 21. *Lexis-Nexis Universe: General News Topics.*

Fox News. (1999, September 25). The Beltway Boys. *Fox News*. Retrieved from *Lexis-Nexis Universe: General News Topics.*

Friedersdorf, C. (2012, October 16). The bizarre story of how drones helped get us into the Iraq war. *The Atlantic*. Retrieved from http://www.theatlantic.com/politics/archive/2012/10/the-bizarre-story-of-how-drones-helped-get-us-into-the-iraq-war/263666/

Gamson, W. A. (1992). *Talking politics*. New York, NY: Cambridge University Press.

Geraghty, T., & Leigh, D. (2002, December 19). Comment & analysis: The name of the game is assassination. *The Guardian*, Guardian Leader, p. 20. *Lexis-Nexis Universe: General News Topics.*

Heath, J. (2001, December 12). Bush sacking ABM treaty; he's expected to give Putin official notice. *Atlanta Journal Constitution*, News, p. A3. *Lexis-Nexis Universe: General News Topics.*

Hutcheson, R. (2001a, December 12). Bush outlines military makeover 3 months after September 11. *Philadelphia Inquirer*, p. A19.

Hutcheson, R. (2001b, December 12). Bush plans major overhaul of military to fight terrorism. *Hamilton Spectator*. Canada & World, C4, *Lexis-Nexis Universe: General News Topics.*

Kagan, D., & Ensor, D. (2002, November 5). Sources: CIA strike kills six terror suspects in Yemen. *CNN Live Today*. *Lexis-Nexis Universe: General News Topics.*

Kaiser, R. G. (2001, December 23). Already, too many pieces are missing. *Washington Post*. Retrieved from https://www.washingtonpost.com/archive/opinions/2001/12/23/already-too-many-pieces-are-missing/90829b44-810d-4f7e-b198-82fc8216ca1a/?utm_term=.a162320e1e55

Lathem, N., & Sujo, A. (2002, November 6). Visit from "hell"—Yemen spies painted bull's-eye on terrorist. *New York Post*, p. 21. *Lexis-Nexis Universe: General News Topics.*

Lobe, J. (2002, November 8). Politics-US: Sovereignty takes major hits in Yemen, Mauritius. *Inter-Press Service*. *Lexis-Nexis Universe: General News Topics.*

MacKenzie, H. (2002, February 12). Pentagon defends fatal CIA missile attack. *The Ottawa Citizen*, News, p. A11. *Lexis-Nexis Universe: General News Topics.*

Mannion, J. (2002, February 11). Allegations of deadly error raise questions about US intelligence. *Agence France Presse-English*. *Lexis-Nexis Universe: General News Topics.*

Martin, D. (2002, November 5). CIA kills key al-Qaida leader. *CBS Morning News*. *Lexis-Nexis Universe: General News Topics.*

Matthews, C. (2002, November 11). Hardball. *MSNBC*. *Lexis-Nexis Universe: General News Topics.*

Michaels, J. (2015). Drones: the face of the war on terror. *USA Today*. Retrieved from http://www.usatoday.com/story/news/world/2015/03/19/drones-pakistan-iraq/25033955/

Miller, J., & Schmitt, E. (2001, November 23). Ugly duckling turns out to be formidable in the air. *New York Times*. Retrieved from http://www.nytimes.com/2001/11/23/international/23PRED.html

Pincus, W. (2002, November 6). Missile strike carried out with Yemeni cooperation: Official says operation authorized under Bush finding. *Washington Post*, A Section, p. A10. *Lexis-Nexis Universe: General News Topics*.

Press Briefing by Ari Fleischer. (2001, December 10). Retrieved from *The American Presidency Project*. Retrieved from http://www.presidency.ucsb.edu/ws/?pid=47613

Press TV. (2015, September 10). Obama ordered ten times more drone strikes than Bush. *Global Research*. Retrieved from http://www.globalresearch.ca/obama-ordered-ten-times-more-drone-strikes-than-bush/5475415

Risen, J., & Johnston, D. (2002, December 14). Bush has widened authority to kill terrorists. *New York Times*, Section 1, p. 1. *Lexis-Nexis Universe: General News Topics*.

Smyth, P. (2001, December 12). Bush plan for military. *The Irish Times*, World News, p. 11. *Lexis-Nexis Universe: General News Topics*.

Sobieraj, S. (2001, December 12). Bush tells cadets terrorism war needs high-tech weapons, old-fashioned spies. *Journal Times*. Retrieved from http://journaltimes.com/news/national/bush-tells-cadets-terrorism-war-needs-high-tech-weapons-old/article_278c4127-2554-56d1-b6d9-36a78d4cce82.html

Solomon, N. (2005). *War made easy: How presidents and pundits keep spinning us to death*. Hoboken, NJ: John Wiley & Sons.

Stone, A. (2002, February 12). Pentagon: Men hit by missile had airline schedules. *USA Today*, News, p. 8A. *Lexis-Nexis Universe: General News Topics*.

Struck, D. (2002, February 11). Casualties of US miscalculations: Afghan victims of CIA missile strike described as peasants, not Al Qaeda. *Washington Post*, A Section, p. A1. *Lexis-Nexis Universe: General News Topics*.

Thomas, E., & Hosenball, M. (2002, November 18). The opening shot. *Newsweek*, International, p. 48. *Lexis-Nexis Universe: General News Topics*.

Whittle, R. (2014). *Predator: The secret origins of the drone revolution*. New York, NY: Henry Holt.

Wilson, S., & Kamen, A. (2009, March 25). "Global war on terror" is given new name. *Washington Post*. Retrieved from http://www.washingtonpost.com/wp-dyn/content/article/2009/03/24/AR2009032402818.html

Wodak, R. (1999). Critical discourse analysis at the end of the 20th century. *Research on Language and Social Interaction*, *32*(1–2), 185–193.

· 3 ·

MURDER INCORPORATED

Turns out I'm really good at killing people.

—Barack Obama discussing the drone program with aides, September 30, 2011[1]

During an October 2, 2011 appearance on CNN's Sunday morning public affairs program, *State of the Union*, former Vice President Dick Cheney praised Barack Obama for the drone strike that killed Muslim cleric Anwar al-Awlaki in Yemen two days earlier. Cheney told reporter Candy Crowley, "I think it was a very good strike. I think it was justified. I think it is very effective use of our drone technology." The former Vice President added, however, that he wanted the Obama administration to apologize for "something they said two years ago when they criticized us for 'overreacting' to the events of 9/11" (Crowley, 2011). Cheney was referring to Obama's 2009 speech in Cairo, where the new president claimed Bush-era counterterrorism tactics violated American values and the rule of law.

With clear satisfaction, the former Vice President suggested Obama has embraced some of those same tactics, adding that Obama's growing reliance on drone strikes demonstrated that his administration "had moved in the direction of taking robust action when they feel it is justified" (Crowley). Having issued an Executive Order revoking Bush's authorization of "enhanced interrogation techniques" within days of taking office, Obama's latitude for taking "robust action"—yet another post-9/11 euphemism for violating domestic and international law—was limited to the covert drone program initiated by his predecessor. This stark realization led former CIA director Leon Panetta to famously describe these killing machines as "the only game in town" for defeating the al-Qaeda network: a widely circulated assessment that served to normalize and institutionalize targeted killing early in Obama's presidency.

Nonetheless, the assassination of Anwar al-Awlaki, an American born citizen and alleged leader of al-Qaeda's Yemen affiliate, set a new precedent in the war on terror: one that had bipartisan support in Congress and full-throated approval from US corporate media, but which met fierce opposition from legal experts, human rights organizations and other critics. The ensuing controversy surrounding the targeting and execution of a US citizen without judicial process constituted a critical discourse moment in America's expanding drone wars. Equally important, al-Awlaki's death serves a link in an intertextual chain that has informed subsequent news stories, policy debates, and critical analyses of the targeted killing program under President Obama (Cohn, 2015; Klaidman, 2012; Scahill, 2013).

Taking the same discourse-historical approach employed in the previous chapter, this discussion charts the ebb and flow of public debate over drone warfare during Obama's presidency. Doing so, I contend the Obama administration's strategy of secrecy and selective leaks was instrumental in shaping news coverage and, ultimately, managing public perception of the US drone program. This chapter proceeds with an examination of news coverage surrounding al-Awlaki's death. Following this, I discuss a series of critical discourse moments in 2013—a watershed in the public life of drones. The chapter concludes with an assessment of the on-again off-again character of public debate over targeted killing. Throughout, I underscore a fundamental characteristic of media discourse—that texts routinely build upon previous ones in a process that either supports and affirms established discursive positions, or disrupts and challenges these positions (Carvalho, 2010). We can detect this discursive struggle in several high profile episodes throughout Obama's tenure as commander-in-chief.

A Major Blow

News accounts of Anwar al-Awlaki employ a range of adjectives—"US-born cleric," "radical Islamist cleric," "Islamic militant cleric," and "American terror cleric"—to characterize the man Obama administration officials classified a "specially designated global terrorist"[2] for his alleged role in the failed 2009 Christmas Day attack over Detroit. These labels reveal a good deal about the "explicit judgments" (Fowler, 1991, p. 110) reporters and editors make in covering the death of suspected terrorists. As Fowler observes, these judgments may not be conscious decisions on the part of news workers; nevertheless, seemingly natural or common sense descriptions reveal ideological perspectives that reflect prejudicial attitudes and sensibilities. Notwithstanding the disparate descriptions for al-Awlaki, media narratives of his death are remarkably consistent in style, tone, and approach.

The majority of news reports echoed and amplified Obama's assessment of al-Awlaki's death, frequently making explicit reference to the President's assertion that the drone strike dealt "a major blow to al-Qaeda's most active operational affiliate" (Viqueira, 2011). So too, numerous press accounts repeated the colorful remarks of an unnamed senior defense official who told reporters, "a very bad man just had a very bad day" (Harnden, 2011). In addition to highlighting the cooperation between US and Yemen intelligence sources in tracking and targeting al-Awlaki, news stories routinely included expert analysis on the consequences this latest assault might have on al-Qaeda's operational, fundraising, and recruitment efforts. Domestic and international reporting also highlighted positive reaction to news of al-Awlaki's death from Obama's Republican opponents in the upcoming presidential election.

For instance, former Massachusetts governor, Mitt Romney, said, "I commend the president, the members of the intelligence community, our service members, and our allies for their continued efforts to keep Americans safe" (Fox News, 2011). Fellow GOP presidential candidates, Texas Governor Rick Perry and former House Speaker Newt Gingrich, expressed likeminded sentiments. And in one of the more effusive statements made by Republicans, Peter King, Chairman of the House Committee on Homeland Security, described al-Awlaki's death as "a tremendous tribute to President Obama and the men and women of our intelligence community" (Fox News, 2011). One notable exception in Republican ranks was presidential hopeful Ron Paul, a libertarian, who called the drone strike an "assassination." While campaigning in New Hampshire, Paul told reporters, "Al Awlaki was born here; he's an American citizen. He was never charged or tried for any crimes." Paul continued, "To start assassinating American citizens without charges—we should think very seriously about this" (Martinez, 2011).

None of which is to suggest that Democrats opposed the drone strike. Following Dick Cheney's appearance on CNN, Jane Harman, former ranking member of the House Intelligence Committee, endorsed the attack, but urged Obama to release a Justice Department memo detailing his administration's justification for killing an American without due process:

> I believe there is a good case, imminent threat beyond our ability to arrest him—the authorization to use military force against al Qaeda. He was complicit with al Qaeda. But I think the Justice Department should release that memo. (Crowley)

But apart from Representative Dennis Kucinich (D-OH), who claimed the Obama administration had "crossed a dangerous divide and set a dangerous precedent for how the United States handles terrorism cases," Congressional Democrats overwhelming supported the President's action (Wilson, 2011).

Al-Awlaki's life story, frequently depicted as something of a puzzle or riddle, is another prominent feature of these accounts. The lead to a CNN report is indicative of the narrative these reports construct: "Born and educated in the United States, how did Anwar al-Awlaki, a charismatic Muslim cleric, go from 'all-American boy' to a terrorist spokesman and recruiter for al Qaeda in the Arabian Peninsula?" (Castillo, 2011). Invariably, these biographies recite a litany of allegations against al-Awlaki: from his role as al-Qaeda's most eloquent and tech-savvy propagandist—reporters sometimes referred to al-Awlaki as the "YouTube Bin Laden"—to his rise as the "leader of external operations" for al-Qaeda in the Arabian Peninsula (AQAP). The subhead "From Inspirational to Operational" used in one wire service report neatly captures this plot device (Hosenball, 2011). Tellingly, these stories repeat administration claims that the cleric was "linked" to al-Qaeda, but fail to acknowledge that al-Awlaki was never formally charged with any crimes.

Finally, print and broadcast reports consistently noted the strike that killed al-Awlaki was one of several attempts to execute him. These stories make clear al-Awlaki had been targeted on at least two previous occasions. Moreover, these accounts describe several FBI investigations into al-Awlaki's relationship with terror suspects and his inclusion, since 2010, on the CIA's "kill list"—making him the first American citizen to be so designated. The attention to al-Awlaki's biography and his radicalization—details of which are purposefully vague in news accounts—stands in sharp relief to the relative paucity of background stories devoted to Samir Khan, a naturalized US citizen of Pakistani heritage and editor of AQAP's online magazine *Inspire*, who perished in the attack on al-Awlaki's caravan. All of which is to suggest that the focus on al-Awlaki's citizenship in subsequent debates rests as much on the cleric's notoriety as it does upon essentialized notions of who is, and who is not, an American citizen.

Amid the triumphalism attending news of al-Awlaki's death, critics condemned the drone strike and warned of its dire implications for constitutional protections at home and humanitarian law abroad. While self-congratulatory story lines dominated news coverage, media organizations did acknowledge the controversy, if not the gravity, of this new precedent in America's decade-long war on terror.

> **Target killing of al-Awlaki raises legal questions**
> (*USA Today*, September 30, 2011)
> **'CELEBRATED' AL-AWLAKI KILLING DISMAYS HUMAN RIGHTS GROUPS**
> (*ISP*, September 30, 2011)
> **LEGAL QUESTIONS SURROUND US ACTION**
> (*Augusta Chronicle*, October 1, 2011)
> **In cleric's killing, an issue of due process; Some question how US can attack American without inquiry or trial**
> (*International Herald Tribune*, October 1, 2011)

Significantly, the sourcing for these stories parallels that of the reports discussed above. That is to say, just as reporters routinely consult official sources in government or the military for comment on US counterterrorism operations, journalists pursuing the legal and constitutional questions raised by al-Awlaki's murder consulted the "usual suspects" in humanitarian and civil liberties organizations—albeit via press releases rather than news conferences or one-on-one interviews. In another instance of intertextuality in news discourse, media organizations reproduced portions of a statement issued by Jameel Jaffer, deputy legal director of the American Civil Liberties Union (ACLU).

> The targeted killing program violates both US and international law. As we've seen today, this is a program under which American citizens far from any battlefield can be executed by their own government without judicial process, and on the basis of standards and evidence that are kept secret not just from the public, but from the courts. (Martinez)

Having offered this dissenting opinion, however, news outlets rarely followed up on the ACLU's allegations that the drone program violates domestic and international law. Similar denunciations by the Center for Constitutional

Rights (CCR) gained even less traction. The Targeted News Service, a wire service for small and mid-market media outlets, was one of the few news organizations to publish CCR Executive Director Vincent Warren's statement, which read in part:

> The targeted assassination program that started under President Bush and expanded under the Obama administration essentially grants the executive the power to kill any US citizen deemed a threat, without any judicial oversight, or any of the rights afforded by our Constitution. If we allow such gross overreaches of power to continue, we are setting the stage for increasing erosions of civil liberties and the rule of law. (Targeted News Service, 2011)

In contrast to the ACLU and CCR's discursive practice of distributing press releases in an effort disseminate legal criticism of the targeted killing program, Johari Abdul-Malik, a member of the Dar al-Hijrah Islamic Center where al-Awlaki preached a decade earlier, declined to take reporters questions, preferring instead to issue a statement. Anxious for reaction to news of al-Awlaki's death at the Northern Virginia mosque, news organizations ran portions of a carefully worded statement[3] that opposed al-Awlaki's killing on religious, as well as legal and constitutional grounds.

> We reiterate that as an American faith community we do not accept violence nor extremism. We have rejected the use of extrajudicial assassination of any human being and especially an American citizen, which includes al-Awlaki. We reiterate our commitment to due process under law and justice, and are concerned that the alleged drone attack sends the wrong message to law-abiding people around the world.

As the statements above demonstrate, a great deal of this criticism focused on questions surrounding al-Awlaki's citizenship; specifically what, if any, legal protections al-Awlaki might enjoy if, as the Obama administration contended, he was actively involved in plotting attacks on Americans. Other critics saw geography, not citizenship, as the crux of the matter. For example, Mary Ellen O'Connell, an expert in international law at the University of Notre Dame, argued that al-Awlaki was "not in an armed conflict zone, not in a battle zone." O'Connell's view contradicts the position staked out two weeks earlier by John Brennan, Obama's counterterrorism chief, who told an audience at Harvard University Law School, "The United States does not view our authority to use military force against al-Qaeda as being restricted solely to 'hot' battlefields like Afghanistan."[4] Nevertheless, O'Connell maintained the cleric should have been arrested, not summarily executed.

> It's basic law, it's basic morality, that you do not kill people without warning in non-battlefield situations. International law is clear about where a battlefield is and is not. We don't get to just make it up because we particularly dislike this guy and we want to declare where he is suddenly a battle zone without doing the hard work of a country to arrest him and put him on trial. (Martinez)

Like other opponents of the drone strike that killed al-Awlaki, O'Connell held a minority opinion: one that was either disparaged or dismissed on the political talk show circuit. On Fox News Channel, Bill O'Reilly referred to critics of al-Awlaki's extrajudicial killing as "far-left pinheads," while his colleague, Geraldo Rivera, suggested groups like the ACLU have "blood on their hands" (O'Reilly & Rivera, 2011). Even more measured commentary reveals impatience with opponents of the drone assassination program. Consider ABC News correspondent Pierre Thomas's dismissal of critics' concerns over judicial process in this appearance on US public television.

> We saw it toward the end of the Bush administration, and ramped up with the Obama administration, the use of drones to kill terrorist targets. They found that it was a very effective way, a very precise way of doing it, and a very clean way of doing it. So they began to say, forget this due process thing in terms of bringing people here, having the national security issues, having trials. If people are clear and present dangers, we're going to kill them. And that's what they've been doing. (Ifill, 2011)

Significantly, Thomas legitimates extrajudicial assassination by evoking the sublime dimensions of drone warfare—its efficiency, precision, and cleanliness—a trope espoused by advocates of targeted killing that animates and legitimates this murderous practice. Doing so, Thomas underscores contrasting approaches to the war on terror under Presidents Bush and Obama. That is to say, while the Bush administration made no qualms over the practices of indefinite detention and enhanced interrogation, Obama opposed these tactics on legal and strategic grounds.

For Obama, revelations of prisoner abuse at Guantánamo and Iraq's Abu Ghraib not only undermined American credibility, but also served as a recruitment tool for extremists. Instead, Obama made clear his preference for the "precision" of the targeted killing program early and often. Writing on the eve of the 2012 Presidential election, Peter Bergen noted, "Covert drone strikes are one of President Obama's key national security policies. He has already authorized 283 strikes in Pakistan, six times more than the number during President George W. Bush's eight years in office" (Bergen & Braun, 2012). Seduced by the sublime capacity to kill by remote control,

Obama and his war planners doubled down on the Bush administration's targeted killing program, while grossly underestimating the blowback this tactic engenders.

All of which is to say, government officials and news workers alike justify the drone assassination program in two ways: first, as a consequence of American technological superiority; and second, as a requirement of American political expediency. Nevertheless, critical perspectives, routinely marginalized in US corporate media, spearheaded renewed debate over America's drone wars. But aside from statements made by civil libertarians, legal scholars and other critical observers, few observers dared utter phrases like "extrajudicial killing" and "violation of international law" to describe the drone program. Still, these dissident voices made a modest, but vitally important contribution to an evolving critique of Obama's drone campaign—one that reached fever pitch in 2013.

Open Secrets and Elastic Definitions

Two weeks after the deaths of Anwar al-Awlaki and Samir Khan, al-Awlaki's 16-year-old son, Abdulrahman, like his father a US citizen, was killed in a drone strike while he ate dinner at an outdoor restaurant in southern Yemen. Six other people perished in an attack that Obama administration officials refuse to explain or defend to this day (Baumann, 2015). In the aftermath of the killing of three Americans in almost as many weeks, lawmakers, news outlets, and human rights organizations demanded greater transparency in Obama's drone campaign, and called on the administration to make public its legal rationale for targeting American citizens without charge or trial.

For the better part of three years, the Obama White House, citing national security interests, stonewalled legal challenges to the secretive drone program, including lawsuits filed by the New York *Times*, the ACLU, and the Center for Constitutional Rights. During that same period, the Obama administration also refused to share the aforementioned Department of Justice (DOJ) memo with relevant Congressional oversight committees. But in February 2013, NBC News obtained a DOJ "white paper"[5] summarizing the administration's legal justification for assassinating American citizens far removed from any declared battlefield. The 16-page document—based on a highly classified memorandum that established the legal framework for placing Anwar al-Awlaki on the CIA's kill list—spurred renewed attention to the targeted killing program and precipitated a short-lived public debate over executive authority, congressional oversight, and judicial review of the increasingly controversial drone campaign.

The leaked document identifies three conditions set out by the Justice Department's Office of Legal Counsel (OLC) that make it lawful for the US government to target American citizens suspected of being "senior operational leaders" of al-Qaeda or "associated forces." First, a "high-level" official must determine that the individual poses an "imminent threat" to US interests. Second, that capture of said individual is "infeasible." And third, that the strike must be conducted in accordance with "law of war principles." The white paper expanded upon earlier public statements by Attorney General Eric Holder[6] and Obama's chief counterterrorism advisor, John Brennan,[7] respectively: the first high-ranking public acknowledgements of the drone program to date. Thus, the white paper revealed for the first time the breadth of the administration's justification for targeting American citizens located far from the battlefield: a legal framework predicated on the president's constitutional responsibility to protect the United States; an admittedly broad interpretation of self-defense; and the Authorization for Use of Military Force (AUMF) passed by the US Congress in the immediate aftermath of the attacks of September 11, 2001.

As Steve Vladeck, professor of law at American University, noted on NPR, successive administrations have used the AUMF to bolster claims of extraordinary executive authority in the conduct of the war on terror: "That was the theory under which we detained prisoners at Guantánamo, and it's the theory behind drone strikes, as well" (Conan, 2013). For instance, the white paper claims the Executive Branch has the authority to assassinate American citizens even in the absence of evidence that the suspect is actively engaged in a terror plot. The memo further stipulates that drone strikes are exempt from judicial review: before and after the fact. Finally, and perhaps most disturbingly, the memo contends that the drone program can operate anywhere in the world.

Discussing the leaked white paper in a rare broadcast interview, the ACLU's Jameel Jaffer claimed the Obama administration is "arguing that the authority to kill American citizens has no geographic limit" (Goodman & Shaikh, 2013). As we shall see, the administration's extraordinary claims led to a dramatic confrontation between the White House and Congress over the President's nominee for Director of Central Intelligence, John Brennan. What's more, the fallout surrounding the leaked DOJ memo politicized the drone campaign as never before; thereby setting the stage for revealing comparisons between Bush-era counterterrorism tactics and those of the Obama administration—a subject taken up at length in Chapter 12.

In addition to the sweeping powers the OLC grants the Executive Branch, the white paper is remarkable for the expansive definitions of "imminence," "self-defense," and "infeasible" used throughout. The document's freewheeling lexical choices did not go unnoticed in news accounts and critical commentary. For instance, Michael Isikoff, the investigative journalist who obtained the leaked document, told MSNBC, "some of the definitions are a bit more *elastic* and open to interpretation than the administration has publicly let on" (emphasis added, Maddow, 2013). Likeminded characterizations appeared frequently in print and broadcast news reports. In an editorial for the New York *Times*, international law expert Mary Ellen O'Connell (2013) argued: "The Justice Department has concocted an elastic definition of necessity, attempting to justify force in the absence of an immediate lethal threat, without citing any treaty or decision by an international court." So too, Hina Shamsi, director of the ACLU's National Security Project, said, "As you read this white paper, you realize that those terms are redefined in such a way that they're vague, elastic, and robbed of their plain meaning" (Hayes, 2013).

In particular, the "elastic" definition of imminence troubled critics. CNN legal analyst Jeffrey Toobin argued the DOJ document evacuates the word of its meaning.

> You know, that's a word that I think we all can understand. It means very soon. It means an hour or maybe a day. This is not imminent. This [white paper] means, well, if you think they might do it sometime in the future. It is such a broad definition of imminent that it essentially seems meaningless to me. (Baldwin, 2013)

Describing the white paper as "chilling," Washington *Post* columnist James Downie echoed this linguistic critique.

> While Webster's defines "imminent" as "ready to take place," the Justice Department has a different take … The memo, rather brilliantly, calls this definition a "broader concept of imminence." Most people, however, would call this "not imminent." Instead, it appears that the Justice Department's version of "imminent" is whatever they say it is. (Downie, 2013)

Expansive and distorted definitions of phrases like "high-ranking officials" and "infeasible" contained in the OLC document were equally suspect; as was the unprecedented authority the white paper granted American authorities. *Esquire* magazine writer, Tom Junod, told CNN:

> It's amazing how much power this white paper gives that informed high level executive. It even really gives him power to decide whether the due process of the person that he's targeting is being violated or not. (Baldwin)

Conversely, the white paper fails to identify precisely which high-ranking officials in the US government can lawfully target individuals for assassination. In light of the secrecy surrounding the drone program, such ambiguity reduces any measure of transparency, let alone public accountability. Furthermore, the document's assertion that drone strikes are a last resort when "capture is infeasible" is unsettling; the white paper indicates that such strikes are preferable when "the relevant country" refuses to sanction or cooperate with ground operations or when such operations would put US forces at "undue risk."[8] Consequently, the word "infeasible" losses its meaning, becoming synonymous with awkward, inopportune and inconvenient. As the ACLU's Jameel Jaffer (2013) concluded, "It's the language of limits—but without any real restrictions."

In sum, the document's vague and self-serving definitions reveal the obfuscatory function of language in legal argumentation generally, and the war on terror in particular (Jackson, 2005). For our purposes, this episode is significant insofar as it represents one of the first instances when news workers interrogated the rather slippery and evasive meanings embedded in the discourse of drone warfare. Such scrutiny was a long time coming. As *Wall Street Journal* Pentagon correspondent, Julian Barnes, noted at the time, drones are "a really important part of American national security policy. It's been a very important weapon in the war on terror. It's an open secret, everyone knows about it." And yet, Barnes concluded, "we just don't know very much about how targets are selected, how the campaign is conducted" (Conan).

#StandWithRand

In light of the twin impediments of official secrecy on the one hand and journalistic acquiescence to the decade-long drone campaign on the other—a subject taken up at length in Chapter 6—the timing of the leak certainly contributed to the unprecedented level of scrutiny the classified white paper received. A year earlier, the New York *Times* (Becker & Shane, 2012) published a report detailing the administration's top secret kill list—including revelations of weekly meetings between Obama and his counterterrorism team dubbed, appropriately enough, "Terror Tuesdays"—but the story failed to gain much traction. However, on the eve of Senate confirmation hearings for John Brennan, President Obama's nominee for CIA director, the white paper's release precipitated a rare filibuster on the floor of the US Senate: a bit of political theater that generated significant news coverage and raised the public profile of the drone assassination program.

This was not the first time Obama tapped Brennan for the CIA leadership position. Five years earlier, concerns over Brennan's involvement in Bush-era detention and interrogation programs scuttled his nomination (Schor, 2008). This time around, Brennan's nomination was considered a fait accompli. But in the wake the DOJ leak, senators from both sides of the aisle publicly expressed concerns over the administration's drone campaign—and the prominent role Brennan played in its escalation and expansion. Days before Brennan's confirmation hearings, a bipartisan group of 11 US Senators sent a letter to the president requesting the release of the classified legal memorandum detailing the rationale for the targeting of American citizens. Months earlier, the White House released the Justice Department white paper to the Senate Intelligence Committee. But now members of the Senate Intelligence, Judiciary, and Armed Services committees were clamoring for the complete DOJ memo.

The showdown between the White House and the US Senate came to a head when freshman Senator Rand Paul (R-KY), son of 2012 Republican presidential candidate Ron Paul, surprised the political and media establishments by launching a filibuster against Brennan's nomination. Rarely used in contemporary political debates, the filibuster is a legislative maneuver—popularized by Jimmy Stewart in the 1939 film *Mr. Smith Goes to Washington*—designed to delay a vote on the Senate floor and allow for the full expression of minority opinions. A case study in intertextuality and interdiscursive practice, Paul's "old-fashioned filibuster" (Friedersdorf, 2013) and the attendant media coverage, reveals the limits and possibilities of opposition to drone warfare in news and official discourse. Equally important, it is indicative of the ebb and flow of public debate over the targeted killing program, as well as the increasingly politicized character of this debate in 2013.

For instance, Senator Paul punctuated his speech with passages from critical commentary on the drone program, including Glen Greenwald's legal analysis in *The Guardian* and Conor Friedersdorf's essay for *The Atlantic* magazine on the death of 16-year-old Abdulrahman al-Awlaki (Mirkinson, 2013). Conversely, news reports alluded to the Hollywood classic, sometimes employing clips and stills of Stewart's dramatic performance, to remind audiences what a filibuster is. In short, the spectacle of the filibuster brought new light to the drone program, prompting *The Nation* magazine's national security correspondent, Jeremy Scahill, to describe Paul's marathon speech "a public service" (Mirkinson).

Senator Paul took the unusual step to voice concern over a letter he received from Attorney General Eric Holder days earlier, which indicated that "under extraordinary circumstances," such as the attacks on Pearl Harbor and September 11th, the Obama administration claimed the authority to target American citizens within the United States (Kludt, 2013). The prospect of targeting Americans on US soil, no matter how remote, prompted Paul to filibuster Brennan's nomination. His opening statement, "I will speak until I can speak no longer" was widely reported. Less so was the following remark:

> I will speak as long as it takes, until the alarm is sounded from coast to coast that our Constitution is important, that your rights to trial by jury are precious, that no American should be killed by a drone on American soil without first being charged with a crime, without first being found to be guilty by a court. (O'Keefe & Blake, 2013)

Here, Paul highlights fundamental questions—the rule of law, the right to due process, and the illegality of summary execution—that were all too infrequent in news and official discourse surrounding the targeted killing program. For the better part of 13 hours, rolling news channels featured live updates of Paul's speech; meanwhile the hashtag #StandWithRand trended on Twitter. Over the course of his discursive marathon, Paul was joined by fellow Republicans Ted Cruz (R-TX) and Marco Rubio (R-FL), as well as Democrat Ron Wyden (D-OR)—one of a handful of critics from the President's own party. Indeed, the dearth of Democratic opposition to Obama's handling of the secretive drone campaign underscored the increasingly politicized nature of the drone debate. Assessing Paul's performance, Peter Beinart of *The Daily Beast* told CNN that the freshman senator was shaming liberals "into saying we have to make sure we don't put our partisanship ahead of our concern about centralized government power and what the Obama administration is doing" (Cooper, 2013).

Meanwhile, conservative news outlets critical of Paul's efforts found it difficult to resist a little word play on his longwinded opposition to Brennan's nomination.

Anti-'CIA' pol drones on and on
(*New York Post*, March 7, 2013)

Rand Paul's Drone Rant
(*Wall Street Journal*, March 7, 2013)

Rand Paul's filibuster—An epic drone about drones
(*Toronto Star*, March 7, 2013)

Off-Target Drone
(*The Economist*, March 7, 2013)

Speaking from the floor of the US Senate the following day, Senator John McCain (R-AZ), no stranger to intertextual practice, castigated Rand Paul and his cohort in remarks that were widely reported.

> Quoting heavily from a Thursday editorial from the *Wall Street Journal* that panned Paul's performance, McCain scolded Paul and other senators who helped him filibuster for doing a disservice to Americans "by making them think that somehow they're in danger from their government." (Douglas & Clark, 2013)

Likewise, Senator Lindsey Graham (R-SC) scarcely concealed his incredulity with US Senators who challenged the president's authority. Doing so, Graham underscored the continuity, and until recently, the bipartisan support for the drone program under both Bush and Obama.

> To my Republican colleagues, I don't remember any of you coming down here suggesting that President Bush was going to kill anybody with a drone, do you? They had a drone program back then, all of a sudden this drone program has gotten every Republican so spun up. What are we up to here? (Troyan, 2013)

Later that day, Eric Holder released a terse letter stating the government no longer claimed the authority to target American citizens on US soil. The letter appeased Paul and his colleagues. By the end of the week, the Senate confirmed Brennan as CIA director. Nevertheless, Paul's political gambit paid off; his filibuster pressured the Obama administration to be more forthcoming in its dealings with congressional oversight committees and, to a lesser extent, the American people.

The episode marked an unprecedented breakdown in elite consensus over the drone program. Elsewhere, the turmoil in Washington prompted members of the European Parliament to weigh in on the increasingly vociferous drone debate. In a statement, parliamentarians said:

> Despite having abandoned the "war on terror" rhetoric, the US sticks to the notion that it is in the realm of a war, and not organised criminality, when fighting terrorism. It has a destabilising effect on the international legal framework …
> We urge our American allies to address the pressing questions over the legal criteria at the basis of a policy that, in targeting so-called militants, destroys both innocent human beings and our common legal heritage. (Crockford, 2013)

In sum, Paul's filibuster called attention to the vexing relationship between technology, international law, and public policy in the war on terror. As Megan Braun (2013), a policy researcher at Oxford University, observed, "Although drone technology is capable of being proportionate and discriminate, the decision makers who use drones do not always exhibit these same traits." Not until Rand Paul drew attention to the (remote) possibility of targeting American citizens on US soil did that decision-making process receive substantive, albeit short-lived, public scrutiny.

Equally important, Paul's filibuster reveals the limits of the drone debate. That is to say, the focus on the *possibility* of American citizens targeted by their own government on US soil disregards far more pressing questions regarding the *reality* of civilian casualties in America's drone wars, let alone the broader civilian impact of robotic warfare. In this respect, news discourse surrounding Rand Paul's filibuster reflects press coverage of drone warfare more generally; news organizations, especially US corporate media, routinely downplay civilian casualties of US drone strikes (Bachman, 2017). Nonetheless, the series of events initiated by the leak of the DOJ white paper and culminating with Senator Paul's filibuster produced a critical discourse moment that opened up discursive space to address the legal, ethical, and moral implications of the decade-long US drone campaign.

By their nature, critical discourse moments of the sort described above ebb and flow over time. As we have seen, the debate over the targeted killing program advances and recedes as events unfold; typically when a breakdown in elite consensus prompts journalists to ask tough questions—a pattern in American news discourse dating back at least to the Vietnam era (Solomon, 2005). Thus, critical discourse moments "are *periods that involve specific happenings*, which may challenge the 'established' discursive positions" (emphasis added, Carvalho, 2010, p. 15). So it is with the debate over drone warfare. Nevertheless, throughout his presidency, Barack Obama successfully navigated episodes that challenged his administration's long-held assertion that the drone program is a precise, legal, and effective means of combatting terrorist threats—thanks, in large measure, to a compliant domestic press corps that dutifully repeated, but rarely interrogated, official statements regarding Obama's drone war.

This routine is upset when American citizens are (potentially) caught in the cross hairs of a drone assault. In response, administration officials launched a public relations campaign designed to minimize press scrutiny and quell public criticism. For instance, in the aftermath of Rand Paul's contentious and widely reported filibuster, the Obama administration set out to reassure Congress and the American people that the drone program was conducted in rigorous and judicious fashion. Speaking on the heels of Senator Paul's filibuster, White House Press Secretary Jay Carney sought to clear up what he characterized as "confusion" surrounding the drone program. Carney told reporters, the President "has not and will not use drone strikes against Americans on American soil," adding, "the issue here is not technology … Whether it's a drone strike or a gun shot, the law and the Constitution apply in the same way" (Knickerbocker, 2013). Time and again, such assurances have been remarkably effective in deflecting criticism and reaffirming the administration's stated positions on the accuracy, precision, and efficiency of drone warfare.

Likewise, when high-ranking officials make public statements regarding the drone program, news reports turn from episodic coverage of drone strikes to more substantive treatment of the legal, political, and ethical implications of drone warfare. For example, on May 23, 2013, President Obama addressed these issues in a highly publicized address at the National Defense University, wherein he laid out Presidential Policy Guidelines (PPG) governing the drone program. The NDU speech is noteworthy for the President's candid appraisal of America's use of military force in the war on terror:

> We cannot use force everywhere that a radical ideology takes root; and in the absence of a strategy that reduces the wellspring of extremism, a perpetual war—through drones or Special Forces or troop deployments—will prove self-defeating, and alter our country in troubling ways.[9]

Following the White House's lead, news reports framed the NDU address as a turning point: one that would recalibrate US counterterrorism strategy for the first time since September 11, 2001. The President not only signaled his willingness to bring greater transparency to the drone campaign, but also promised to renew efforts to close the detention facility at Guantánamo, and to work with Congress to "refine, and ultimately repeal"[10] the 2001 Authorization for Use of Military Force (AUMF). Obama's speech generated significant coverage, but little in the way of follow up reporting. In the absence of sustained public scrutiny of US counterterrorism strategy, a year to the day following the President's address, Human Rights Watch's Andrea Prasow (2014) wryly noted, "none of these promises have been met."

In June 2014, a federal appeals court in New York City ordered the release the controversial DOJ memo. While civil liberties groups praised the document's release as an important step toward ensuring greater transparency in the drone program, the highly redacted document provided few new details on the Obama administration's legal justification of targeting American citizens. More critically, the release of the DOJ memo and its implications for challenging official secrecy failed to generate much traction beyond the 24/7 news cycle. Notwithstanding Senator Ron Wyden's contention that the release of the long-secret memo might "generate new pressure for the executive branch to answer other pressing questions," (Landay, 2014) critical questions regarding the targeting of alleged terrorists and reporting procedures on civilian causalities in America's drone wars remained largely unanswered.

The following year, two Western hostages held by al-Qaeda on the Afghanistan-Pakistan border were killed in a US drone strike. On April 23, 2015, President Obama issued an unprecedented public apology for the attack that "inadvertently" killed American aid worker Warren Weinstein and Giovanni Lo Porto, an Italian citizen. While news outlets noted the rarity of the President's regrets, few reporters saw fit to address fundamental flaws in the targeted killing program this incident exposed. The New York *Times* muted reaction, headlined "Drone Strikes Reveal Uncomfortable Truth: US is Often Unsure About Who Will Die," is indicative of journalists' timid and all too infrequent critique of America's drone wars (Shane, 2015). Thus, this latest incident renewed concerns over the targeted killing program. But like the uproar surrounding al-Awlaki's extrajudicial assassination four years earlier, the debate was short-lived. In the meantime, the US drone campaign continued apace.

***

When he assumed office in January 2009, Barack Obama inherited two clandestine programs from his predecessor George W. Bush: the CIA's "enhanced interrogation" program and the targeted killing campaign (Zenko, 2016). On his second day in office, Obama banned torture. And the very next day, the new president ordered two drone strikes in northwest Pakistan reported to have killed one militant—and as many as 10 civilians. For Micah Zenko, a senior fellow at the Council for Foreign Relations, the escalation and expansion of the US drone program constitutes Obama's lasting foreign policy legacy: a legacy that raises troubling questions about how the United States identifies and targets suspected terrorists. The flipside of this foreign policy is the prospect of increased hostility toward America and its allies—a recipe for what Obama aptly described as "perpetual war."

On the domestic front, Zenko concludes that Obama has made targeted killing acceptable to the American people: a public relations coup based as much on the administration's discursive strategies of secrecy, selective leaks, and obfuscation, as it is upon a complacent, and frequently complicit, domestic press corps. Indeed, in the weeks following Obama's apology for the death of two Western hostages, an Associated Press-GfK poll found a majority of Americans continue to support the drone program—a sharp contrast to overwhelming opposition to weaponized drones in world opinion (Fuller, 2014). Revealingly, the American people are far more ambivalent about the prospect of commercial and civilian drones operating in domestic airspace—one more indication of the negative lust drone technology provokes in the public imagination. For entrepreneurs, investors, and drone enthusiasts eager to fly unmanned vehicles for fun and profit, legal and regulatory obstacles abound; and yet, as we shall see, the domestication of this new technology represents an even more formidable cultural challenge: overcoming the public's fear and apprehension of these flying machines. It is to this discursive struggle to domesticate drones that we now turn.

Notes

1. See Halpern, M., & Heilemann, J. (2013). *Double down: Game change 2012* (p. 55). New York, NY: Penguin Press.
2. This information appears primarily in international accounts of al-Awlaki's death. For example, see "Anwar al-Awlaki" http://www.telegraph.co.uk/news/obituaries/8800037/Anwar-al-Awlaki.html.
3. See "Daj Al-Hijrah Islamic Center Statement," https://www.onfaith.co/onfaith/2011/09/30/dar-al-hijrah-islamic-center-statement-on-death-of-anwar-al-aulaqi/11108.
4. See "Strengthening our Security by Adhering to our Values and Laws," https://www.whitehouse.gov/the-press-office/2011/09/16/remarks-john-o-brennan-strengthening-our-security-adhering-our-values-an.
5. See "DOJ White Paper," http://msnbcmedia.msn.com/i/msnbc/sections/news/020413_DOJ_White_Paper.pdf.
6. See "Attorney General Eric Holder Speaks at Northwestern University School of Law," http://www.justice.gov/opa/speech/attorney-general-eric-holder-speaks-northwestern-university-school-law.

7. See "Brennan's Speech on Counterterrorism," http://www.cfr.org/counterterrorism/brennans-speech-counterterrorism-april-2012/p28100.
8. See "DOJ White Paper," http://msnbcmedia.msn.com/i/msnbc/sections/news/020413_DOJ_White_Paper.pdf.
9. See "Remarks by the President at the National Defense University," https://www.whitehouse.gov/the-press-office/2013/05/23/remarks-president-national-defense-university.
10. *Ibid.*

References

Bachman, J. (2017). The New York Times and Washington Post: Misleading the public about US drone strikes. *Journalism Studies, 18*(4), 470–494.

Baldwin, B. (2013, February 5). CNN Newsroom. *CNN. Lexis-Nexus Universe: General News Topics.*

Baumann, N. (2015, April 23). The American teen whose death-by-drone Obama won't explain. *Mother Jones.* Retrieved from http://www.motherjones.com/kevin-drum/2015/04/abdulrahman-al-awlaki-obama-drone

Becker, J., & Shane, S. (2012, May 29). Secret "kill list" tests Obama's principles. *New York Times.* Retrieved from http://www.nytimes.com/2012/05/29/world/obamas-leadership-in-war-on-al-qaeda.html

Bergen, P., & Braun, M. (2012, September 19). Drone is Obama's weapon of choice. *CNN.com.* Retrieved from http://www.cnn.com/2012/09/05/opinion/bergen-obama-drone/

Braun, M. (2013, March 7). Where Rand Paul and John Brennan can agree on US drone program. *Christian Science Monitor.* Retrieved from http://www.csmonitor.com/Commentary/Opinion/2013/0307/Where-Rand-Paul-and-John-Brennan-can-agree-on-US-drone-program-video

Carvalho, A. (2010). Media(ted) discourse and society: Rethinking the framework of critical discourse analysis. In J. Richardson (Ed.), *Language and journalism* (pp. 10–26). New York, NY: Routledge.

Castillo, M. (2011, September 30). American-born radical cleric understood the west. *CNN.com.* Retrieved from http://www.cnn.com/2011/09/30/world/meast/yemen-al-awlaki-profile/

Cohn, M. (Ed.). (2015). *Drones and targeted killing: Legal, moral, and geopolitical issues.* Northampton, MA: Olive Branch Press.

Conan, N. (2013, February 7). A preview of Brennan's confirmation hearings. *NPR.* Retrieved from http://www.npr.org/2013/02/07/171392872/a-preview-of-brennans-confirmation-hearings

Cooper, A. (2013, March 6). Anderson Cooper 360 degrees. *CNN. Lexis-Nexis Universe: General News Topics.*

Crockford, K. (2013, March 7). Obama must reveal legal memos on his administration's drone killing program. *The Guardian.* Retrieved from https://www.theguardian.com/commentisfree/2013/mar/07/obama-reveal-legal-memos-drone-killing

Crowley, C. (2011, October 2). State of the Union with Candy Crowley. *CNN.com.* Retrieved from http://transcripts.cnn.com/TRANSCRIPTS/1110/02/sotu.01.html

Douglas, W., & Clark, L. (2013, March 7). Senate confirms CIA chief—after GOP senators rip Rand Paul for filibuster. *McClatchy*. Retrieved from http://www.mcclatchydc.com/news/politics-government/congress/article24746242.html

Downie, J. (2013, February 5). The justice department's chilling "targeted killings" memo. *Washington Post*. Retrieved from https://www.washingtonpost.com/blogs/post-partisan/wp/2013/02/05/justice-department-chilling-drone-white-paper/?utm_term=.911e298e0e02

Fowler, R. (1991). *Language in the news: Discourse and ideology in the press*. London: Routledge.

Fox News. (2011, September 30). Washington praises killing of al-Awlaki. *Fox News*. Retrieved from http://www.foxnews.com/politics/2011/09/30/washington-praises-killing-al-awlaki.html

Friedersdorf, C. (2013, March 7). Cliff notes for the filibuster: Rand Paul in his own words. *The Atlantic*. Retrieved from https://www.theatlantic.com/politics/archive/2013/03/cliffs-notes-for-the-filibuster-rand-paul-in-his-own-words/273787/

Fuller, J. (2014, July 15). Americans are fine with drone strikes. Everyone else in the world? Not so much. *Washington Post*. Retrieved from https://www.washingtonpost.com/news/the-fix/wp/2014/07/15/americans-are-fine-with-drone-strikes-everyone-else-in-the-world-not-so-much/?utm_term=.542e9fe91bc8

Goodman, A., & Shaikh, N. (2013, February 5). Kill list exposed: Leaked memo shows assassination of US citizens "has no geographic limit." *Democracy Now!* Retrieved from https://www.democracynow.org/2013/2/5/kill_list_exposed_leaked_obama_memo

Harnden, T. (2011, October 1). America hails "a very bad day for a very bad man"; the United States. *The Daily Telegraph*, National Edition, p. 18. *Lexis-Nexis Universe: General News Topics*.

Hayes, C. (2013, February 10). Up with Chris Hayes. *MSNBC*. *Lexis-Nexis Universe: General News Topics*.

Hosenball, M. (2011, October 7). Secret panel can put Americans on "kill list." *Reuters*. Retrieved from http://www.reuters.com/article/us-cia-killlist-idUSTRE79475C20111005

Ifill, G. (2011, September 30). PBS's Washington Week with Gwen Ifill. *Federal News Service*. *Lexis-Nexis Universe: General News Topics*.

Jackson, R. (2005). *Writing the war on terrorism: Language, politics and counter-terrorism*. Manchester: Manchester University Press.

Jaffer, J. (2013, February 4). The justice department's white paper on targeted killing. *ACLU*. Retrieved from https://www.aclu.org/blog/justice-departments-white-paper-targeted-killing

Klaidman, D. (2012). *Kill or capture: The war on terror and the soul of the Obama presidency*. New York, NY: Houghton Mifflin.

Kludt, T. (2013, March 6). Holder allows for drone strikes in US under "extraordinary circumstance." *Talking Points Memo*. Retrieved from http://talkingpointsmemo.com/livewire/holder-allows-for-drone-strikes-in-u-s-under-extraordinary-circumstance

Knickerbocker, B. (2013, March 7). Rand Paul's drone filibuster shakes up Republicans. *Christian Science Monitor*. Retrieved from http://www.csmonitor.com/USA/Politics/DC-Decoder/2013/0307/Rand-Paul-s-drone-filibuster-shakes-up-Republicans

Landay, J. S. (2014, June 23). Justice department releases controversial memo on legality of drone strike on American. *McClatchy*. Retrieved from http://www.mcclatchydc.com/news/crime/article24769603.html

Maddow, R. (2013, February 4). The Rachel Maddow Show. *MSNBC*. *Lexis-Nexis Universe: General News Topics*.

Martinez, M. (2011, September 30). US drone killing of American al-Awlaki prompts legal, moral debate. *CNNPolitics.com*. Retrieved from http://www.cnn.com/2011/09/30/politics/targeting-us-citizens/

Mirkinson, J. (2013, March 6). Rand Paul cites a range of media sources during filibuster of John Brennan. *Huffington Post*. Retrieved from http://www.huffingtonpost.com/2013/03/06/rand-paul-media-filibuster_n_2822704.html

O'Connell, M. E. (2013, February 6). The questions Brennan can't dodge. *New York Times*. Retrieved from http://www.nytimes.com/2013/02/07/opinion/the-questions-brennan-cant-dodge.html

O'Keefe, E., & Blake, A. (2013, March 6). Rand Paul launches talking filibuster against John Brennan. *Washington Post*. Retrieved from https://www.washingtonpost.com/news/post-politics/wp/2013/03/06/rand-paul-begins-talking-filibuster-against-john-brennan/?utm_term=.b7b58f808e2d

O'Reilly, B., & Rivera, G. (2011, September 30). The O'Reilly Factor. *Fox News*. *Lexis-Nexis Universe: General News Topics*.

Prasow, A. (2014, May 23). The year of living dangerously: Obama's drone speech was a sham. *The Guardian*. Retrieved from http://www.theguardian.com/commentisfree/2014/may/23/obama-drone-speech-one-year-later

Scahill, J. (2013, June 7). Inside the US dirty war in Yemen with Jeremy Scahill, Nasser al-Awlaki, Sheikh Fareed. *Democracy Now!* Retrieved from https://www.democracynow.org/2013/6/7/inside_the_us_dirty_war_in

Schor, E. (2008, November 25). Leading candidate to head CIA under Obama withdraws from running. *The Guardian*. Retrieved from https://www.theguardian.com/world/2008/nov/25/obama-white-house-cia-brennan

Shane, S. (2015, April 23). Drone strikes reveal uncomfortable truth: US is often unsure about who will die. *New York Times*. Retrieved from http://www.nytimes.com/2015/04/24/world/asia/drone-strikes-reveal-uncomfortable-truth-us-is-often-unsure-about-who-will-die.html

Solomon, N. (2005). *War made easy: How presidents and pundits keep spinning us to death*. Hoboken, NJ: John Wiley & Sons.

Targeted News Service. (2011, September 30). CCR condemns targeted assassination of US citizen Anwar Al-Awlaki. *Targeted News Service*. *Lexis-Nexis Universal: General News Topics*.

Troyan, M. O. (2013, March 7). Senators bash fellow Republican for comments on drones. *USA Today*. Retrieved from http://www.usatoday.com/story/news/politics/2013/03/07/paul-bashed-for-drone-comments/1971575/

Viqueira, M. (2011, October 1). Drone strike kills high-ranking al-Qaeda members. *NBC News*, Saturday Today. *Lexis-Nexis Universe: General News Topics*.

Wilson, S. (2011, September 30). Obama pushes boundaries in targeting al-Aulaqi. *Washington Post*. Retrieved from https://www.washingtonpost.com/national/national-security/obama-pushes-boundaries-in-targeting-al-aulaqi/2011/09/30/gIQATZrFBL_story.html?utm_term=.7ae13fa91925

Zenko, M. (2016, January 12). Obama's embrace of drone strikes will be a lasting legacy. *New York Times*. Retrieved from http://www.nytimes.com/roomfordebate/2016/01/12/reflecting-on-obamas-presidency/obamas-embrace-of-drone-strikes-will-be-a-lasting-legacy

PART II

DOMESTICATING DRONES

· 4 ·

UNMANNED

Drones for Fun and Profit

> *For once you have tasted flight you will walk the earth with your eyes turned skywards, for there you have been and there you will long to return.*
>
> —Leonardo da Vinci

In the early morning hours of January 26, 2015, an errant DJI Phantom quadcopter—what the digital magazine of science, technology and culture, *The Verge*, affectionately describes as "the Kleenex of drones"—crashed on the White House lawn (Popper, 2015). According to published reports, security personnel "heard and observed" the aircraft "flying at a very low altitude" near the White House sometime after 3 AM, triggering a facility-wide lockdown (Jackson & Jansen, 2015). Later that day, the Secret Service issued a statement indicating that the operator reported the loss of the hobby aircraft to authorities. The investigation found no evidence of malicious intent and concluded the incident was simply the result of "recreational use of the device" (Fox News, 2015).

The latest in a series of security breaches around the Executive residence, the story made national headlines. Not least for the episode's unintentional irony. For starters, the pilot, who admitted to authorities he was drinking at the time of the accident, works for the National Geospatial-Intelligence

Agency (Shear & Schmidt, 2015). And the following day, Barack Obama, who oversaw the escalation and expansion of America's clandestine drone program, proposed greater government oversight of this new class of flying machine. Speaking with CNN, the President argued that a lack of clear operating guidelines undermined the commercial and civilian potential of the new technology. Calling on the Federal Aviation Administration (FAA) and other agencies to expedite the integration of unmanned aircraft into domestic airspace, the President concluded: "There are incredibly useful functions that these drones can play … but we don't really have any kind of regulatory structure at all for it" (Diamond, 2015). As the New York *Times*' opinion page editor wryly noted: "The same could be said for Mr. Obama's drone assassination program" (Lapidos, 2015).

This observation underscores a central dynamic in a second order of drone discourse—what I refer to throughout this section as a *discourse of domestication*—surrounding the use of this new technology for commercial, law enforcement, public safety, and a host of civilian applications. That is, proponents invariably qualify their enthusiasm for drones operating in domestic airspace by implicitly acknowledging the lethality, if not the moral ambiguity, of robotic warfare. From manufacturers and investors to photographers and conservationists, drone advocates employ a variety of discursive strategies to ease cultural anxieties produced by a dominant order of discourse that casts drones as sublime killing machines—harbingers of a new kind of war. To that end, industry representatives, marketers, and enthusiasts celebrate the awe-inspiring versatility of these flying machines to save lives and generate profits; create new jobs and remake entire industries; protect endangered species and fragile ecosystems; and offer new realms of sensorial experience. Like public relations efforts designed to domesticate nuclear energy following America's use of the atomic bomb at the end of World War II, this domesticating discourse seeks to dissociate drones from their military applications; offering instead a new understanding of drones as a beneficial and revolutionary technology: a boon to industry, commerce, and the arts; sublime objects of excitement, fascination, and wonder.

This chapter first considers the importance of this domesticating discourse for destigmatizing drones for an anxious American people. Here, a discursive strategy associating drones with more familiar instances of technological innovation, such as the personal computer and the cell phone, plays a decisive role in facilitating public acceptance of these flying machines. Appeals to American entrepreneurialism and technological ingenuity likewise offer new ways of thinking about and understanding drones. Not as menacing tools of death and destruction, but rather as profit making opportunities, labor saving devices, and leisure time pursuits (Niehenke, 2016). In other words, this discourse of domestication offers an altogether different perspective on the new technology: one that highlights the social, economic, and political benefits of unmanned vehicles (Boucher, 2015; Choi-Fitzpatrick, 2014). Throughout, I highlight discursive strategies and practices designed to demystify the technology and overcome public apprehension of these flying machines.

Following this, I consider drones in the realm of entertainment and popular culture—like news discourse (Chapter 6), an important site for the domestication of unmanned aerial vehicles. Significantly, drones are not just another fearsome technology—like Hal 9000, the sentient supercomputer of Stanley Kubrick's *2001: A Space Odyssey*—to serve as a plot device in literature, theater, film and television. Rather, photographers, filmmakers, and even graffiti artists have embraced drones to produce their art (Aquino, 2015; Cheng, 2016; Vanhemert, 2014). As noted in Chapter 1, the technology's aesthetic potential plays a decisive role in constructing public knowledge of, and fueling cultural fascination with, these flying machines. Accordingly, Arthur Holland Michel (2015) contends, "by summarizing and crystallizing popular feelings and assumptions about drones, these representations are a useful barometer for understanding how the public understands the drone, and what it thinks about it." The chapter concludes with a brief discussion of recent efforts to develop new modes of expression and engagement with drone technology. Here, I note the cultural affinity between this quintessentially twenty-first century technological form and the illuminated spectacles of the late nineteenth century that helped domesticate the technological sublime of the day: the electric light.

Airborne Enterprise

Founded in 1972, the Association for Unmanned Vehicle Systems International (AUVSI) bills itself as "the world's largest nonprofit organization devoted exclusively to advancing the unmanned systems and robotics community" (AUVSI, n.d.). For most of its history, AUVSI labored in relative obscurity promoting military and civilian uses of remotely piloted vehicles, later rechristened unmanned systems, capable of operating on land, air, and sea. That all began to change in the 1980s, when the aerospace industry, in collaboration with the US Defense Advanced Research Projects Agency (DARPA), began realizing the potential unmanned systems held for surveillance, reconnaissance, and tactical operations (Zaloga, 2008). In the wake of the attacks of September 11, 2001, the Bush administration embarked on a policy of targeted assassination using the most successful UAV at the time, General Atomics' Predator, to track and kill al Qaeda leadership (Hersh, 2002). As detailed in Chapters 2 and 3, drones quickly became synonymous with America's war on terror, raising the technology's public profile and stimulating a new wave of research and development. "In less than a decade, military drone technology proliferated to an extent that even its most ardent advocates never imagined, and an explosion in civilian drone technology followed" (Whittle, 2014a, p. 299).

AUVSI set out to capitalize on this sudden interest in unmanned systems. In early 2013, the trade organization released an economic forecast of civilian drone integration in the United States through the year 2025. Widely reported in the business and trade press, the study projects the creation of more than 100,000 new jobs, with an overall economic impact of $82 billion (AUVSI, 2013). Notwithstanding criticism of the study's methodology and findings,[1] AUVSI's figures, like Amazon's highly publicized plans for drone delivery, constitute an intertextual chain in subsequent news accounts extolling the economic benefits of drones (Dillow, 2013; Ellman, 2017; Pramuk, 2015). AUVSI has since led the charge to integrate unmanned vehicles into domestic airspace: a formidable regulatory challenge made all the more difficult in light of public ambivalence toward the technology. Hence AUVSI's short-lived campaign, described in the book's introduction, to eliminate the word drone—and its indelible association with targeted killing—from popular usage. By December 2014, AUVSI realized it was fighting a losing discursive battle and reversed course, enthusiastically embracing the "D-word" in press conferences and promotional material; most notably its Know Before You Fly[2] campaign aimed at educating hobbyists of the FAA's rules for operating recreational drones ahead of the Christmas buying season (Whittle, 2014b).

Speaking with the Center for the Study of the Drone (CSD) in January 2015, outgoing AUVSI president Michael Toscano articulated the industry's position on civilian and commercial applications of drone technology: one that acknowledges public apprehension over the military, public safety, and privacy implications of the new technology, but which emphasizes the social

Figure 4.1: The DJI Phantom takes flight at the 2015 Consumer Electronics Show. *Photo Credit: Maurizio Pesce. Source: Creative Commons.*

and economic benefits of unmanned systems. While the general public may be most familiar with the dramatic aerial perspectives photographers acquire through unmanned systems, increasing productivity in the agricultural sector is a favored talking point for industry representatives. According to Toscano, "Our study was predicated on the idea that agriculture would be a huge industry" for commercial drones (CSD, 2015). Indeed, the technology's principal applications—as delivery and aerial imaging systems—are uniquely suited to crop management. Farmers use unmanned systems to conserve resources and maximize yields by delivering water, pesticides, and fertilizer to crops (Main, 2013). So-called precision agriculture further increases productivity by giving farmers the ability to survey their fields in a more time and cost-effective fashion (Stangler, 2013). As one landowner explained in *sUAS News*, "I just view a UAV as another tool that should be sitting in your shed somewhere that comes out when it's the appropriate time, to help maximize your farming productivity" (Press, 2014).

This discourse of domestication employs a two-fold strategy: one that echoes news and official discourse hailing drones as an efficient and cost-effective weapon system, while demilitarizing unmanned vehicles by way of analogy with more familiar technological forms. For instance, Toscano likens the rise of civilian UAVs to the rapid adoption of cellular telephony:

> No one could have envisioned where this technology has gone. It's almost like the cell phone in the year 2000. Two percent of the world had a cell phone. And here we are 15 years later and more than 80% of the world has a cell phone. (CSD, 2015)

Toscano's argument, that what was once an unwieldy and frequently unreliable communication technology is now ubiquitous, aims to demystify drones while alleviating fears raised by the prospect of UAVs sharing America's crowded airspace with commercial and passenger aircraft. "It's all about safety," Toscano adds reassuringly. And risk acceptance. Toscano offers the automobile as a case in point. Recounting the logistical problems associated with integrating the car into complex urban environments in the early twentieth century, Toscano describes the automobile's enormous benefits—as well its drawbacks. "Here in the United States we have over 33,000 deaths per year. … And yet, we still drive cars every day because the upside is so great that we are willing to accept the downside of it" (CSD). Likening UAVs with more commonplace machines in this fashion, Toscano normalizes domestic applications of drone technology.

Central to realizing this objective, Toscano promotes the use of microprocessors—components routinely found in smartphones and all manner of GPS-enabled devices—for real-time "sense and avoid" guidance systems that reduce the risk of pilot error and mid-air collisions. The development of such systems, what Toscano refers to as "the holy grail" of domestic drone integration, heralds a boon to commercial, civilian, and recreational applications (Figure 4.1). While hailing their revolutionary potential, Toscano is quick to underscore the evolutionary character of these flying machines. "What you've utilized is the technology that has developed from cell phones and from different other aspects of technology and just put it together" (CSD). These arguments echo those of Chris Anderson, editor-in-chief of *Wired* magazine and founder of DIY Drones, a social network for recreational drone enthusiasts. Anderson writes:

> This new generation of cheap, small drones is essentially a fleet of flying smartphones. More and more, autopilot electronics look just like smartphone electronics, simply running different software. The technical and economic advantages of coat tailing on the economies of scale of the trillion-dollar mobile-phone industry are astounding. If you want to understand why the personal-drone revolution is happening now, look no farther than your pocket. (2012)

Anderson effectively recasts a strange and futuristic technology as something altogether safe, highly profitable, and instantly recognizable: the smartphone. Toscano takes a similar approach to address privacy concerns raised by unmanned vehicles. Here, the analogy is with the personal computer, and the costs and benefits associated with the integration of information technology into daily life.

> The same thing with computers: the computers help us, but it helps the bad guys. And the bad side with computers is you've got identity theft, pornography, bullying—you've got things that people didn't intend the technology to ever be used for. But bad people will use technology to do bad things. And I believe you have to hold people accountable. (CSD)

On this final point, Toscano gestures toward the intersection between technological innovation and public policy, but fails to pursue this line of thought. It is a self-serving and short-sighted approach to technological development; not unlike the one taken by US war planners, whose deployment of armed drones sets troubling precedents for global security, international relations, and the rule of law (Falk, 2015). Rather, Toscano's more immediate aim, to *dissociate* drones from their military applications, is paramount. Equating drones with more familiar (and seemingly indispensible) technologies such as the cell phone, the automobile, and the personal computer, serves to demystify drones through association with devices that were likewise met with initial skepticism, if not outright resistance. Doing so, AUVSI pursues the twin strategic objectives of demilitarizing unmanned vehicles and naturalizing the technology for domestic purposes.

For Chris Anderson, the user-driven evolution of the personal computer and the advent of networked communication offer additional clues for thinking about drone integration. Recalling the military origins of the internet, Anderson argues that everyday people appropriated the new technology for their own purposes. "The amateur UAV community is hoping to do the same with drones, demilitarize and democratize them so they can find their full potential" (Anderson, 2012). Praising the DIY sensibility of drone enthusiasts, Anderson contends legions of altruistic hobbyists will liberate the technology from its more ominous applications and "set it free." Here, Anderson echoes the rhetoric of the technological sublime discussed in Chapter 1, wherein innovations in manufacturing, communication, and transportation are routinely associated with democratic ideals and aspirations. Equally important, linking drones with more familiar technologies lends an air of inevitability to the domestic integration of unmanned vehicles.

For proponents, the real impediment to domestic drone flight is not public apprehension, but rather an unnecessarily complicated and restrictive regulatory regime that stymies technological development and economic growth (Poletti, 2016). According to AUVSI's Toscano, once "normal people can see how it can affect their life, you will see the numbers like I mentioned with the cell phone go exponential" (CSD). From an industry perspective, then, the ability to field the technology is essential for realizing its potential and easing public anxiety. In the days following the quadcopter crash on the White House lawn, technology reporter Farhad Manjoo (2015) lamented what he called "another in a string of public relations disasters" for the drone industry. Nevertheless, Manjoo contends the incident was not, as some observers suggested, indicative of an impending public safety crisis. Instead, Manjoo argues that public understanding of the new technology has been distorted by burdensome government regulations that restrain business innovation: "If the rules are loosened and commercial operations take flight, drone makers argue that perceptions will shift."

This discursive strategy focused on demilitarization, economic development, and technological innovation eventually won the day. Following years of industry lobbying, government testing, and policy analysis, in June 2016, the FAA cleared the way for commercial drone operation.[3] The regulations, which went into effect in August 2016, do not require commercial operators to obtain a pilot's license—according to proponents, regulations requiring such a license was a major stumbling block to drone integration. As a result, a diverse group of would-be drone operators, from filmmakers and realtors to farmers and environmentalists, celebrated the new regulations (Grady, 2016). Taking their cues from industry representatives, journalists described the FAA decision in terms of overcoming "barriers" and clearing "obstacles" to commercial drone flight. Moreover, the aforementioned AUVSI estimates on job creation and economic growth figured prominently in both trade and popular press accounts (e.g., Anon, 2016; Maney, 2016; Traverse, 2013). Although the guidelines prohibit drone delivery schemes, like Amazon's highly anticipated Prime Air service, they open the door to a host of commercial and civilian applications. For financial analyst Troy Jensen, the FAA's long awaited guidelines have enormous consequences for manufacturers and investors alike: "With favorable regulation now in place, we believe the commercial drone industry in on track to become one of the better secular growth industries and we advise investors to gain exposure to the space" (Seitz, 2016).

Even before the FAA regulations went into effect, financiers saw drones as a unique investment opportunity. In 2014, when the FAA announced the establishment of six national test sites for commercial drone operation, news accounts echoed and amplified industry talking points promoting drones as engines of job creation and economic development (e.g., Wozniacka, 2014). Notwithstanding manufacturers' frequently hyperbolic rhetoric, the drone economy was clearly taking shape. For instance, the complex regulatory environment surrounding civilian drone operation prompted law firms to develop expertise in unmanned vehicles. Citing the increasingly dense web of federal, state, and local regulations governing domestic drone flight, *Fortune* magazine concluded:

> Given the myriad applications for the technology, the scope of "drone law" is broad, stretching from freedom of speech and press concerns to issues around intellectual property or land use rights, as well as more practical matters of insurance and liability. (Dillow, 2014)

Predictably, the prospect of unmanned vehicles filling the skies led to the development of "counter-drone" technologies. These defensive systems represent yet another growth industry (Pavluk & Cole, 2016). For instance, NoFlyZone lets homeowners designate their properties as a drone-free area. Upon registration to a global database, NoFlyZone creates a virtual barrier that protects homeowners' privacy. According to CEO Ben Marcus: "We already have the technology to create geofence exclusion zones like the one around the White House, but now we're expanding that technology to private property" (Wheeler, 2015). More recently, the San Francisco-based technology company, Dedrone, raised $15 million in investment capital to develop systems that can detect and disarm rouge drones. Dedrone deployed such a system at several high profile events, including the World Economic Forum in Davos and the 2016 Presidential debate in Las Vegas (Glaser, 2017). Amidst news reports of jihadist groups weaponizing recreational drones to attack their adversaries, calls for ever more sophisticated anti-drone technologies represent additional investment opportunities (Mortimer, 2017).

Meanwhile, as manufacturers and investors sought to domesticate drones for commercial and civilian use, the market for military drones grew exponentially. In September 2010, the Los Angeles *Times* reported on the "buzz" surrounding California's aerospace industry. The story describes lucrative defense contracts secured by General Atomics, manufacturer of the Predator and Reaper aircraft that have come to epitomize US counterterrorism strategy, in glowing terms:

> The company's 1.9-million-square-foot facility is a showcase for Southern California's drone industry, which employs an estimated 10,000 people. The fast-growing business is fueled by Pentagon spending—at least $20 billion since 2001—and billions more chipped in by the CIA and Congress. (Hennigan, 2010)

Five years later, Jane's Intelligence Review, a leading authority on weapons procurement and proliferation, released a study indicating the worldwide market for military drones would exceed $10 billion by 2024. Citing increased competition in a fast-growing drone manufacturing sector, the report notes that in 2015 the United States outperformed Israel, the onetime leader in military UAV sales. Looking ahead, defense analyst Huw Williams foresees a new generation of unmanned systems featuring stealth technology and increasingly sophisticated armaments. "They will operate alongside manned aircraft and eventually even replace them in many roles" (Agence France Presse, 2015).

From high altitude surveillance platforms to swarming semi-autonomous vehicles, drones have irrevocably changed the face of modern warfare. And with each passing day, all manner of unmanned vehicles operate in domestic airspace. Small wonder, then, that drones have so thoroughly captured the public imagination. The sublime attraction—and repulsion—toward these flying machines is manifest, most vividly perhaps, in popular entertainment.

Trigger Warnings

Season 4 of the award-winning comedy, *Arrested Development*, finds man-child Buster Bluth re-enlisting in the military as a drone pilot. Under the false impression he is playing a first-person shooter game, a clueless Buster revels in his killing prowess. When it dawns on him that he actually blew up an art museum, in Spain no less, an apoplectic Buster tumbles out of his armchair, becoming the first active duty drone operator to sustain bodily injuries. As Jesse Kirkpatrick (2016) contends, lampooning remote pilots in this fashion minimizes the emotional, cognitive, and psychological harm that comes from participating in America's targeted killing campaign. Nevertheless, this satirical take on drone warfare is not uncommon—a subject taken up at length in Chapter 12. Rather, it is emblematic of the technology's spectacular ascent in popular culture. Foregoing fine-grained analysis of a single text for a more comprehensive survey, I briefly examine the depiction of drones, and drone pilots, across three genres: action adventure films, political thrillers, and psychological dramas.

The figure of the superhero has proven to be an extraordinarily malleable character in American popular culture, uniquely suited to address changing socio-political forces and conditions (Hatfield, Heer, & Worcester, 2013). Accordingly, weaponized drones make one of their first big screen appearances in *Iron Man 2* (Feige & Favreau, 2010). Blending Cold War rivalries with war on terror technology, *Iron Man 2* finds billionaire weapons designer, Tony Stark/Iron Man (Robert Downey, Jr.), battling Russian physicist Ivan Vanko (Mickey Rourke). Vanko has co-opted Stark's design for Iron Man's armored suit and created a fleet of semi-autonomous weapons, so-called Hammer drones. For Vanko, modern warfare is best left to the machines: "People are the problem," he snarls. Like their real world counterparts, the robots are unveiled with great fanfare at a military-technology expo. Vanko then unleashes the Hammer drones on the unsuspecting crowd. As Stark fights the robots—tailored to meet the specific needs of each branch of the US armed forces—the inhumanity of drone warfare plays out in allegorical fashion against a backdrop of mechanized mayhem: a staple of action adventure films. The climax of this effects-intensive sequence comes when a Hammer drone sets its merciless sights on a young boy. Stark arrives just in time to save the child from certain death. Thus, Stark protects the innocent amid the chaos and confusion of robotic warfare. As we shall see, cultural producers frequently deploy this ideologically charged plot device.

Another action adventure film, *Star Trek: Into Darkness* (Burk & Abrams, 2013), finds the crew of the USS Enterprise ordered to Qo'noS, the homeworld of the Klingon Empire, to kill John Harrison, a rogue Starfleet officer responsible for a series of terrorist attacks. Using advanced long-range torpedoes, the Enterprise will execute Harrison without violating Klingon territory, thereby avoiding interstellar war. Enroute Mr. Spock (Zachary Quinto) questions Starfleet's orders, arguing that Harrison be captured and tried for his crimes. Captain Kirk (Chris Pine) agrees and leads a landing party down to the planet surface to arrest Harrison (Benedict Cumberbatch). In keeping with the franchise's penchant for displacing contemporary political issues in time and space (Franklin, 1994), this scenario reflects several aspects of America's war on terror. For instance, Harrison's refuge, a desolate and sparsely populated province, stands in for remote parts of the globe, such as Pakistan's

Federally Administered Tribal Areas (FATA), where the United States has conducted its most intense drone campaign (Coll, 2014). And of course the torpedoes that enable the Enterprise to target an individual for assassination from vast distances call to mind America's globe-spanning drone program. Finally, Kirk's decision to arrest, rather than assassinate, Harrison encapsulates the central tension in US counterterrorism strategy: to capture or kill suspected terrorists (Jaffer, 2016; Zenko, 2015). Ironically then, Captain Kirk, always one to play fast and loose with the rules, adheres to the principles of due process and apprehends the fugitive. Doing so, *Star Trek: Into Darkness* evokes the legal and political implications of the US drone program in the guise of a sci-fi adventure movie.

The most overtly political critique of drone warfare in the action adventure genre, *Captain America: The Winter Soldier* (Feige, Russo, & Russo 2014), is set in the present day. Combining blockbuster aesthetics with elements of the political thriller, the film depicts the US national security state run amok. Under the leadership of Alexander Pierce (Robert Redford) the security agency S.H.I.E.L.D. launches Project Insight: a weaponized surveillance system featuring three enormous killer drones, Helicarriers, ostensibly intended to promote peace and security. In reality, Helicarriers conduct preemptive strikes on anyone the state deems a security risk. Designed to remain airborne indefinitely, Helicarriers employ data mining algorithms to identify and track individuals who *might* become future threats. A brief but revealing exchange between Captain America (Chris Evans) and Nick Fury (Samuel L. Jackson) underscores the film's preoccupation with preemptive war.

NICK FURY: We're going to neutralize a lot of threats before they even happen.

CAPTAIN AMERICA: I thought the punishment usually came after the crime.

Questions of crime and punishment are perennial concerns of the action adventure genre. How well these films answer those questions is, of course, another matter (Davidson, 2013). Still, this installment of the Marvel Comic superhero series is not without merit. While exaggerating the firepower and surveillance capability of remotely piloted aircraft (RPAs) currently in use, the film underscores the increasingly prominent role signals intelligence (SIGINT) plays in conducting drone strikes—more about this in Chapter 9 and the concluding chapter. Equally important, the film reflects concerns over successive administrations' use of preemptive strikes to eliminate individuals who may, or may not, pose an "imminent threat" to US interests. Constrained by the conventions of the action adventure film, *Captain America: The Winter Soldier* nevertheless draws upon and dramatizes cultural anxieties associated with targeted killing, government surveillance, and rampant abuses of the national security state.

Television producers have addressed similar concerns in a number of critically acclaimed thrillers. Part of an increasingly pervasive tendency in post 9/11 culture, these programs mobilize the "national security imagination" (Kumar & Kundnani, 2014) to normalize America's decades-long war on terror. Fueled by an unnerving mix of xenophobia, jingoism, and paranoia, two distinct television series epitomize this trend: *24* (Grazer & Gordon, 2001–2010) and *Homeland* (Gansa & Gordon, 2011–). In the first season of *Homeland*, CIA agent Carrie Mathison (Claire Danes), suspects Nicholas Brody (Damian Lewis), an Iraq War veteran held prisoner by al-Qaeda for 8 years, of being a mole for the extremist group. During his captivity Brody befriends his captor and converts to Islam. When a US drone strike kills Issa, a young boy Brody had been tutoring, the marine vows revenge on his own government. The death of an innocent child caught up in a global conflict puts the action in motion.

Mathison pursues her quarry with obsessive determination. Herein lies an underlying affinity between ostensibly different characters: *Homeland's* shrewd heroine, Carrie Mathison, and *24*'s shoot-first-and-ask-questions-later hero, Jack Bauer (Keifer Sutherland). Unlike Bauer, however, whose personal suffering barely slows him down, Mathison's traumatic combat experience frequently clouds her judgment. Complicating matters, Mathison conceals her bi-polar disorder from colleagues and superiors, further jeopardizing her own safety and theirs. Throughout the series, Mathison proves her mettle in the field, and later, as CIA station chief in Afghanistan, where colleagues dub her "The Drone Queen." Nonetheless, in the figure of Carrie Mathison, this post 9/11 heroine is dangerously unstable.

The source of a great deal of the program's narrative tension, Mathison's erratic behavior and emotional instability amplifies and reinforces unflattering gender stereotypes. Likewise, *24: Live Another Day* (Grazer & Gordon, 2014) features an obsessive female antagonist, Margot Al-Harazi (Michelle Fairley), hell-bent on exacting revenge for her husband's death. Implausibly, Margot has commandeered six American drones and plots to assassinate US President James Heller (William Devane) while he is in London negotiating a deal to keep a controversial military base open. Railing against the US drone strike that killed her husband, along with dozens of innocent children, Margot demands the President turn himself in. If not, she will unleash an aerial assault on London. True to form, Jack Bauer saves Heller in a nick of time and repels the drone attack. Reveling in the unconventional, frequently illegal, and decidedly immoral tactics employed by Bauer and Mathison, these programs perform the important ideological work of naturalizing US counterterrorism strategy (Kumar & Kundnani). Conversely, both *Homeland* and *24: Live Another Day* depict emotionally unstable women, with dubious motives, wielding weaponized drones. Doing so, both programs subtly but unmistakably reassert the need for levelheaded patriarchal authority in the struggle against Islamic extremism: a defining feature of 9/11 culture (Melnick, 2009).

Gender politics, evident in a subtle critique of Islamic fundamentalist laws governing women's lives, provide a subtext for *Eye in the Sky* (Doherty & Hood, 2015). The military thriller pivots on the capture or kill conundrum raised by *Star Trek: Into Darkness*. Indeed, many of the film's themes and plot devices resonate with our previous examples. For instance, like *24: Live Another Day*, *Eye in the Sky* unfolds in real time: a narrative device that heightens tension and draws viewers into the action. And while the drones in *Captain America: The Winter Soldier* are outsized versions of contemporary RPAs, *Eye in the Sky* anticipates the use of micro-RPAs to surveil suspected terrorists in close quarters. Furthermore, like *Iron Man 2* and *Homeland*, when a child gets caught in the cross hairs of a US drone strike midway through *Eye in the Sky*, the plot takes an unexpected twist. Anger and remorse over the death of innocent children likewise underpins the trauma drone pilots experience in the psychological dramas *Grounded* and *Good Kill*.

All told, like its predecessors, *Eye in the Sky* offers an upbeat assessment of the technological infrastructure of targeted assassination: facial recognition software to identify terror suspects; computer modeling programs to generate collateral damage estimates; and digital communication systems to coordinate drone strikes (Currier, 2016). What makes the film distinctive is its dramatization and critique of the "kill chain" (Cockburn, 2015): a globally dispersed network of remote pilots, sensor operators, intelligence analysts, lawyers, military and political leaders who authorize, coordinate, and conduct drone strikes. Exploring the tension between legal obligations, political expediency, and military prerogatives, the film yields productive insights into the geo-politics of targeted killing. Released within days of an unprecedented US-led assault that claimed 150 lives at an alleged Al-Shabab training camp in Somalia (Cooper, 2015), *Eye in the Sky* underscores the dearth of public debate over the ethics of targeted killing; and its consequences for the service men and women who pull the trigger, not to mention the innocent civilians on the receiving end of Hellfire missiles launched from American drones.

While *Eye in the Sky* places remote pilots in the middle of a sprawling bureaucracy, the multi-media stage production *Grounded* (Brant, 2013) and feature film *Good Kill* (Amin & Niccol, 2014) focus squarely on the psychological and emotional toll robotic warfare exacts from drone operators. In *Grounded*, an unnamed female F-16 fighter pilot, recently reassigned to flying RPAs, tells her story in an extended monologue. George Brant's character study gained national attention when Oscar-winning actress Anne Hathaway teamed up with Tony-winning director Julie Taymor to stage a production at New York's Public Theater. Grounded after learning she is pregnant, the pilot reluctantly returns to duty as part of what she derisively refers to as the "Chair Force." Her longing to return to "the blue" is clear, as is her initial contempt for piloting remote vehicles. Nevertheless, with a loving family waiting for her at the end of the day, she begins to appreciate her new lifestyle as a remote control assassin. "I will see my daughter grow up. I will kiss my husband good night. … The threat of death has been removed." Over time, the sublime appeal of killing with impunity is displaced by the cognitive dissonance of the daily commute to a virtual battlefield—where the human carnage is palpable and all too real.

Both *Grounded* and *Good Kill* examine the repercussions of what Melissa Gregg (2011) refers to as "presence bleed": an Information Age condition that finds the demands of work impinging on personal relationships and responsibilities. For drone operators, the secrecy surrounding the drone program, coupled with the unnerving intimacy produced by robotic warfare, exacerbates presence bleed, leading to alienation and despair. Despite a script burdened with exposition of the technology and logistics of drone warfare, *Good Kill* offers a disturbing portrait of a former F-16 pilot coming to terms with this new kind of war. The film reaches its dramatic climax when Major Thomas Eagan (Ethan Hawke) appoints himself judge, jury, and executioner. Over the course of the film, Eagan's crew witnesses several brutal sexual assaults of an Afghan woman. Prohibited from intervening, the crew's frustration is obvious. One airman reminds his fellow crewmembers, "He's a bad guy. He's just not our bad guy." Increasingly estranged from his wife and family, and self-medicating with booze, Eagan takes it upon himself to dispense justice by unleashing a Hellfire missile and ending the serial rapist's life. The film's conclusion reaffirms a common, if altogether specious justification for US military intervention across the Muslim world: defending women and girls against the oppression and savagery of Islamic culture (Jabbra, 2006). Notwithstanding the ethical objections to drone warfare raised throughout the film, *Good Kill* ultimately legitimates the practice of extrajudicial assassination.

Days of Future Past

On March 23, 2013 Paramount Pictures took to the skies above London to promote the upcoming release of *Star Trek: Into Darkness*. Timed to coincide with the World Wildlife Federation's annual "Earth Hour" event, which encourages municipalities across the globe to conserve energy by dimming non-essential lights, 30 illuminated quadcopters took flight in the darkened skies to form a rotating, 300 foot-long Starfleet insignia adjacent to the city's iconic Tower Bridge (Griner, 2013). In dramatic fashion, the publicity stunt leverages the franchise's mythos of social progress and technological wonderment with leading-edge aerial robotics. More important for my purposes, evoking *Star Trek's* twenty-third century technological sublime, the event harkens to a bygone era when industrialists, electrical engineers, and promoters took to the skies with illuminated spectacles that would usher in the era of electric communication. As media historian Carolyn Marvin (1988) observes, these future-oriented light shows "were easily accessible and enthusiastically relished by their nineteenth-century audiences" (p. 153). With increasing frequency, and growing popularity, drone manufactures, IT companies, and marketers employ much the same strategy as their nineteenth-century counterparts to domesticate unmanned systems through ever more sophisticated public displays.

Intel, the computer chip manufacturer, has emerged as a leader in these twenty-first century illuminated spectacles. Using its trademarked Shooting Star drones, Intel is at the vanguard of choreographed light shows featuring hundreds of illuminated quadcopters. Partnering with a constellation of corporate conglomerates, including Disney and the National Football League (NFL), Intel works with brands to develop new forms of consumer spectacles and new modes of storytelling. For instance, in November 2016, Intel's Shooting Star drones took center stage at Disney's seasonal spectacular, "Starlight Holidays." According to a press release: "It marks the first time in the US that 300 drones simultaneously perform a synchronized light show. The flying robots paint the sky with colorful images choreographed to traditional holiday music" (Kaplan, 2016). Grafting new forms of communication on more traditional public celebrations, like Disney's holiday festivities, echoes the application of electric lights on more "familiar spectacles" during the late nineteenth century (Marvin, p. 153). Furthermore, Intel's collaboration with Disney yields productive synergies for two disparate corporations whose combined expertise produces innovative forms of immersive storytelling.

Intel's Shooting Star drones also shared the spotlight with Lady Gaga during the 2017 Super Bowl halftime show. Once again, there were corporate synergies aplenty. An estimated 160 million television viewers witnessed Intel's dancing drones form an American flag at the opening of the grandiosely titled Pepsi Zero Sugar Super Bowl LI Halftime Show. As it turns out, the drone choreography behind Lady Gaga's rendition of "America the Beautiful" was pre-recorded for safety reasons, in compliance with FAA regulations governing the operation of drones above crowded stadiums (Barrett, 2017). Nevertheless, the light show's use of corporate logos and the American flag underscore yet another relationship between Intel's drone-powered light shows and its nineteenth century antecedents. That is, in some its earliest incarnations, electric communication was used for purposes of business promotion—a lesson Intel has rebooted for the new century. Furthermore, electric light shows proved indispensible for celebrating national holidays and promoting a shared sense of patriotism. In this way, public spectacles domesticated the new technology through dramatic appeals to national unity and identity. Intel's contribution to the Super Bowl halftime show resonates with these earlier public displays of electric communication.

Finally, Marvin contends the illuminated spectacles of a bygone era persisted in the dominant mass medium of the twentieth century: "In modern television, the element of the spectacle recalls the electric light show, the most dramatic tradition of electrical experimentation in the late nineteenth century" (p. 157). Drones have likewise made inroads into television. In April 2016, the International Drone Racing League (DRL) announced its partnership with ESPN, the sports network, to televise the US National Drone Racing championships (Marshall, 2016). Part video game, part real-world sporting event, drone pilots don virtual reality (VR) goggles to fly DRL Racers through as series of obstacles. Video from on-board cameras allows spectators to experience the action with extraordinary intimacy (Olewitz, 2016). The event was live streamed on ESPN-3 and subsequently edited for broadcast. Exploiting the dramatic first-person-view (FPV) afforded by drone racers and replete with spectacular crashes—not to mention ample opportunity for both virtual and embedded advertising—Drone Racing League CEO Nick Horbaczwski hopes to turn drone racing into the next NASCAR (NPR, 2016). Like other forms of popular entertainment, then, this twenty-first century synthesis of technology and spectacle, sports and marketing, evokes the technological sublime in the service of naturalizing drones for domestic consumption.

Notes

1. See "Five reasons the AUVSI got its drone market forecast wrong," *sUAS News*, https://www.suasnews.com/2014/06/five-reasons-the-auvsi-got-its-drone-market-forecast-wrong/.
2. AUVSI conducts the ongoing educational campaign with its partners: the FAA, the Academy of Model Aeronautics, and the Small UAV Coalition. See http://knowbeforeyoufly.org.
3. For a concise review of the regulatory process surrounding domestic integration of unmanned vehicles, see Rupprecht (2015).

References

Agence France Presse. (2015, October 3). *Experts: Drone market to hit $10 billion in 2024*. Retrieved from http://www.defensenews.com/story/defense/air-space/2015/10/03/experts-drone-market-hit-10-billion-2024/73282590/

Amin, N., (Producer), & Niccol, A. (Director). (2014). *Good Kill*. [Motion picture]. United States: Voltage Pictures.

Anderson, C. (2012, July 26). Here come the drones! *Wired*. Retrieved from http://www.wired.co.uk/article/here-come-the-drones

Anon. (2016, October). New USA FAA rules clear drones for industrial use. *Environmental Engineering*, *29*(5), 4.

Aquino, T. (2015, December 15). How drone technology is changing Hollywood. *Complex*. Retrieved from http://www.complex.com/pop-culture/2015/12/how-drones-are-changing-up-the-entertainment-industry

AUVSI. (n.d.). Who is AUVSI. *Association for Unmanned Vehicle Systems International*. Retrieved from http://www.auvsi.org/home/learnmore

AUVSI. (2013). The economic impact of unmanned aircraft systems in the United States. *Association for Unmanned Vehicle Systems International*. Retrieved from https://higherlogicdownload.s3.amazonaws.com/AUVSI/958c920a-7f9b-4ad2-9807-f9a4e95d1ef1/UploadedImages/New_Economic%20Report%202013%20Full.pdf

Barrett, B. (2017, February 5). Lady Gaga's super bowl LI halftime show drones have a bright future. *Wired*. Retrieved from https://www.wired.com/2017/02/lady-gaga-halftime-show-drones/

Boucher, P. (2015). Domesticating the drone: The demilitarisation of unmanned aircraft for civil markets. *Science & Engineering Ethics*, *21*(6), 1393–1412.

Brant, G. (2013). *Grounded*. London: Oberon Books.

Burk, B., (Producer), & Abrams, J. J. (Director). (2013). *Star Trek. Into Darkness*. [Motion picture]. United States: Paramount Pictures.

Center for the Study of the Drone. (2015, January 22). *Interview: Michael Toscano*. Retrieved from http://dronecenter.bard.edu/interview-michael-toscano/

Cheng, E. (2016). *Aerial photography and videography using drones*. Berkeley, CA: Peachpit Press.

Choi-Fitzpatrick, A. (2014). Drones for good: Technological innovations, social movements, and the state. *Journal of International Affairs*, 68(1), 19–36.

Cockburn, A. (2015). *Kill chain: The rise of the high-tech assassins*. New York, NY: Henry Holt.

Coll, S. (2014, November 24). The unblinking stare. *The New Yorker*. Retrieved from http://www.newyorker.com/magazine/2014/11/24/unblinking-stare

Cooper, H. (2015, March 7). US strikes in Somalia kill 150 Shabab fighters. *New York Times*. Retrieved from https://www.nytimes.com/2016/03/08/world/africa/us-airstrikes-somalia.html?_r=0

Currier, C. (2016, March 19). Drone warfare's ethical dilemmas are focus of film "Eye in the Sky." *The Intercept*. Retrieved from https://theintercept.com/2016/03/19/drone-warfares-ethical-dilemmas-are-focus-of-film-eye-in-the-sky/

Davidson, A. (2013, May 20). Is "Star Trek Into Darkness" a drone allegory? *New Yorker*. Retrieved from http://www.newyorker.com/news/amy-davidson/is-star-trek-into-darkness-a-drone-allegory

Diamond, J. (2015, January 28). Obama: We need more drone regulations. *CNNPolitics.com*. Retrieved from http://www.cnn.com/2015/01/27/politics/obama-drones-fareed/index.html

Dillow, C. (2013, May 13). Inside the drone economy. *Fortune*. Retrieved from http://fortune.com/2013/05/13/inside-the-drone-economy

Dillow, C. (2014, April 9). "Drone law" becomes big business. *Fortune*. Retrieved from http://fortune.com/2014/04/09/drone-law-becomes-big-business/

Doherty, G., (Producer), & Hood, G. (Director). (2015). *Eye in the Sky*. [Motion picture]. United Kingdom: Entertainment One.

Ellman, L. (2017, January 2). 3 ways the Trump administration can help the drone industry take off. *The Hill*. Retrieved from http://thehill.com/blogs/pundits-blog/technology/312147-3-ways-the-trump-administration-can-help-the-drone-industry

Falk, R. (2015). Why drones are more dangerous than nuclear weapons. In M. Cohn (Ed.), *Drones and targeted killing: Legal, moral, and geopolitical issues* (pp. 29–49). Northampton, MA: Olive Branch Press.

Feige, K., (Producer), & Favreau, J. (Director). (2010). *Iron Man 2*. [Motion picture]. United States: Paramount Pictures.

Feige, K., (Producer), Russo, A., & Russo, J. (Directors). (2014). *Captain America: The Winter Soldier*. [Motion picture]. United States: Marvel Entertainment.

Fox News. (2015, January 26). Man claims responsibility for drone crash at White House, says was an accident. *Fox News*. Retrieved from http://www.foxnews.com/politics/2015/01/26/obama-spokesman-says-device-found-on-white-house-grounds-poses-no-threat.html

Franklin, H. B. (1994). *Star Trek* in the Vietnam era. *Science Fiction Studies, 21*(1), 24–34.

Gansa, A., & Gordon, H. (Producers). (2011-). *Homeland.* [Television series]. Hollywood, CA: Showtime.

Glaser, A. (2017, February 14). A company that detects and disarms drones engaging in illegal activity just raised $15 million. *Recode.* Retrieved from http://www.recode.net/2017/2/14/14600030/dedrone-disarm-illegal-drones-15-million-funding

Grady, D. (2016, December 17). The unmanned frontier: Drone entrepreneurs take advantage of new FAA rule. *News & Tribune*. Retrieved from http://www.newsandtribune.com/news/the-unmanned-frontier-drone-entrepreneurs-take-advantage-of-new-faa/article_a71bb-bcc-c3cc-11e6-9464-83bd7caf8d30.html

Grazer, B., & Gordon, H. (Producers). (2001–2010). *24.* [Television series]. Hollywood, CA: 20th Century Fox.

Grazer, B., & Gordon, H. (Producers). (2014). *24: Live Another Day.* [Television mini-series]. Hollywood, CA: 20th Century Fox.

Gregg, M. (2011). *Work's intimacy.* Cambridge: Polity Press.

Griner, D. (2013, March 25). Robot copters hovering over London form Starfleet logo to promote next *Star Trek* film. *Adweek.* Retrieved from http://www.adweek.com/creativity/robot-copters-hovering-over-london-form-starfleet-logo-promote-next-star-trek-film-148151/

Hatfield, C., Heer, J., & Worcester, K. (Eds.). (2013). *The superhero reader.* Jackson, MS: University Press of Mississippi.

Hennigan, W. J. (2010, September 12). Drones create a buzz in Southern California aerospace industry. *Los Angeles Times.* Retrieved from http://articles.latimes.com/2010/sep/12/business/la-fi-drones-20100912

Hersh, S. (2002, December 23). Manhunt. *The New Yorker.* Retrieved from http://archives.newyorker.com/?i=2002-12-23#folio=066

Holland Michel, A. (2015, September 15). Drones in popular culture. *Center for the Study of the Drone.* Retrieved from http://dronecenter.bard.edu/drones-popular-culture/

Jabbra, N. W. (2006). Women, words and war: Explaining 9/11 and justifying US military action in Afghanistan and Iraq. *Journal of International Women's Studies*, 8(1), 236–255.

Jackson, D., & Jansen, B. (2015, January 26). Government employee piloted drone that crashed at White House. *USA Today.* Retrieved from http://www.usatoday.com/story/news/politics/2015/01/26/obama-white-house-device-josh-earnest/22338781/

Jaffer, J. (2016). *The drone memos: Targeted killing, secrecy and the law.* New York, NY: The New Press.

Kaplan, K. (2016, November 16). Disney makes history with first US light show powered by 300 Intel drones. *iQ by Intel.* Retrieved from http://iq.intel.com/drone-light-show-takes-disney-world-visitors-future/

Kirkpatrick, J. (2016, January 26). Military drone operators risk a serious injury. *Slate.com.* Retrieved from http://www.slate.com/articles/technology/future_tense/2016/01/military_drone_operators_risk_moral_injury.html

Kumar, D., & Kundnani, A. (2014). Imagining national security: The CIA, Hollywood, and the war on terror. *Democratic Communiqué*, 26(2), 72–83.

Lapidos, J. (2015, January 27). Drones at the White House. And in Yemen. *New York Times*. Retrieved from https://takingnote.blogs.nytimes.com/2015/01/27/drones-at-the-white-house-and-in-yemen/?action=click&contentCollection=U.S.&module=RelatedCoverage&region=Marginalia&pgtype=article

Main, D. (2013, September 12). Yes, there's such a thing as a good drone, and this is what it does. *Takepart*. Retrieved from http://www.takepart.com/article/2013/09/11/uses-drones-we-can-all-get-behind

Maney, K. (2016, November 16). Drones unleashed. *Newsweek*, pp. 46–47.

Manjoo, F. (2015, February 4). Giving drone industry leeway to innovate. *New York Times*. Retrieved from https://www.nytimes.com/2015/02/05/technology/personaltech/giving-the-drone-industry-the-leeway-to-innovate.html?_r=0

Marshall, A. (2016, April 15). ESPN has decided drone racing is a sport because internet. *Wired*. Retrieved from https://www.wired.com/2016/04/espn-decided-drone-racing-sport-internet

Marvin, C. (1988). *When old technologies were new: Thinking about electric communication in the nineteenth century*. New York, NY: Oxford University Press.

Melnick, J. (2009). *9/11 culture*. Malden, MA: Blackwell.

Mortimer, A. (2017, January 5). Anti drone measures a growing market. *sUAS News*. Retrieved from https://www.suasnews.com/2017/01/anti-drone-measures-growing-market/

Niehenke, A. (2016, November 13). How drones will reshape the enterprise. *TechCrunch*. Retrieved from https://techcrunch.com/2016/11/13/how-drones-will-reshape-the-enterprise/

NPR. (2016, January 31). A video game IRL: Drone Racing League aims to be NASCAR in the air. *All Tech Considered*. Retrieved from http://www.npr.org/sections/alltechconsidered/2016/01/31/464714071/a-video-game-irl-drone-racing-league-aims-to-be-nascar-in-the-air

Olewitz, C. (2016, January 29). Competitive Drone Racing League launches 2016 season. *Digital Trends*. Retrieved from http://www.digitaltrends.com/cool-tech/competitive-drone-racing-league-launches-2016-season/

Pavluk, J., & Cole, A. (2016, February 23). Four counter-drone technologies we need now. *TechCrunch*. Retrieved from https://techcrunch.com/2016/02/23/four-counter-drone-technologies-we-need-now/

Poletti, T. (2016, April 5). Drone startups fly high, but regulatory concerns are a drag. *MarketWatch*. Retrieved from http://www.marketwatch.com/story/drone-startups-fly-high-but-regulatory-concerns-are-a-drag-2016-04-05

Popper, B. (2015, January 26). Our favorite drone crash-landed at the White House. *The Verge*. Retrieved from http://www.theverge.com/2015/1/26/7917119/drone-crash-white-house-dji-phantom

Pramuk, J. (2015, September 20). Lethal drones are the industry's latest headache. *CNBC*. Retrieved from http://www.cnbc.com/2015/09/18/lethal-drones-are-the-industrys-latest-headache

Press. (2014, May 29). Upper Midwest farmers employ drone technology. *sUAS News*. Retrieved from https://www.suasnews.com/2014/05/upper-midwest-farmers-employ-drone-technology

Rupprecht, J. (2015). *Drones: Their many civilian uses and the US laws surrounding them*. Palm Beach Gardens, FL: Author.

Seitz, P. (2016, June 24). Commercial drone market cleared for takeoff by FAA. *Investor's Business Daily*. Retrieved from http://www.investors.com/news/technology/click/commercial-drone-market-cleared-for-takeoff-by-faa/

Shear, M. D., & Schmidt, M. S. (2015, January 27). White House drone crash described as a US worker's drunken lark. *New York Times*. Retrieved from https://www.nytimes.com/2015/01/28/us/white-house-drone.html?_r=0

Stangler, C. (2013, August 16). The drone boom. *In These Times*. Retrieved from http://inthesetimes.com/article/15476/the_drone_boom

Traverse, N. (2013, May 13). Idea of the week: The drone economy. *New Yorker*. Retrieved from http://www.newyorker.com/news/news-desk/idea-of-the-week-the-drone-economy

Vanhemert, K. (2014, April 16). This open source graffiti drone will give cops nightmares. *Wired*. Retrieved from https://www.wired.com/2014/04/a-nightmare-for-cops-a-drone-that-paints-graffiti/

Wheeler, L. (2015, February 10). Company launches no-drone database for private residents. *The Hill*. Retrieved from http://thehill.com/regulation/232298-company-launches-no-fly-drone-zone-database-for-private-residents

Whittle, R. (2014a). *Predator: The secret origins of the drone revolution*. New York, NY: Henry Holt.

Whittle, R. (2014b, December 23). Drone defender drops D-word denial. *Breaking Defense*. Retrieved from http://breakingdefense.com/2014/12/drone-defender-drops-d-word-denial

Wozniacka, G. (2014, January 1). Drone test sites spell opportunity for companies. *Seattle Times*. Retrieved from http://www.seattletimes.com/seattle-news/drone-test-sites-spell-opportunity-for-companies

Zaloga, S. J. (2008). *Unmanned aerial vehicles: Robotic air warfare 1917–2007*. Oxford: Osprey Publishing.

Zenko, M. (2015, April 14). Kill > capture. *Foreign Policy*. Retrieved from http://foreignpolicy.com/2015/04/14/kill-capture-obama-drone-pakistan-cia-policy-special-operations

· 5 ·

EYE IN THE SKY

New Surveillance Regimes

> *We are rapidly entering the age of no privacy, where everyone is open to surveillance at all times; where there are no secrets from government.*
>
> —Supreme Court Justice William O. Douglas, *Osborn v. United States*, 1966

A virtual news blackout surrounding the months-long Dakota Access Pipeline protests ended abruptly on October 23, 2016, when authorities from the Morton County Sheriff's Department shot down a UAV activists had been using to document the growing resistance movement. In a widely circulated statement, Sheriff Kyle Kirchmeier claimed the drone approached a police helicopter "in a threatening manner," adding that the pilot and passengers were "in fear for their lives" (CBS/AP, 2016). Over the course of the nonviolent protest against pipeline construction under the Missouri river, the primary source of drinking water for the Standing Rock Sioux reservation, self-described "water protectors" deployed the aircraft to record their encounters with local law enforcement; live streaming video to Facebook and posting footage to YouTube.

Video showing police using rubber bullets, percussion grenades, and water cannons against activists garnered thousands of viewers and proved instrumental in enlisting the support of indigenous and non-native people across

North America (Koebler, 2016a). Despite compelling evidence of local authorities using military technology and tactics to assault water protectors, corporate news outlets ignored the story for weeks. But when police shot down an unmanned vehicle, the incident made headlines (Ellis & Grinberg, 2016; Zilber, 2016). Such is the news value of sensational stories, like those discussed in Chapter 1, involving "rogue" drones: a sharp contrast to mainstream media silence surrounding the militarization of domestic law enforcement, state repression of nonviolent protest, and violations of press freedoms unfolding at Standing Rock (Listening Post, 2016).

Like their predecessors in the Occupy Wall Street and Black Lives Matter movements, NoDAPL activists leveraged consumer electronics with social media platforms to document mass demonstrations, raise public awareness, and hold local authorities to account.

> Early on in the protests, outside attention was captured by birds-eye views of the camps and the wound cut through the North Dakotan landscape for the pipeline. As the protests continued and police forces ratcheted up tactics to end the protests … the drones caught footage of police violence against peaceful protestors in the same way the #BlackLivesMatter movement captured video of police violence against African American communities. (Herzog, 2016)

According to Myron Dewey, whose indigenous media startup, Digital Smoke Signals, had been documenting the resistance encampment since August 2016, drones are nothing short of a game changer. "For years, the story has been told for us, not by us. This is our opportunity to tell our story" (Koebler, 2016a). As noted in Chapter 4, drones are unique storytelling tools. In this instance, UAVs enable citizen journalists to tell the story of local law enforcement employing aggressive, frequently unconstitutional tactics to suppress nonviolent resistance. Doing so, protestors upend state surveillance regimes with a form of counter-surveillance scholars refer to as *synopticism*: thereby turning the observed into the observer, the policed into the policer (Sandhu & Haggerty, 2017). Another indigenous journalist, Shiye Bidziil, who posts footage of the NoDAPL resistance to the Facebook page Drone2bwild, echoes this sentiment: "Simple drone technology has given us activists and grassroots people the upper hand in surveillance" (Herzog). Whether or not activists can maintain the upper hand is, of course, another question (Kopstein, 2017).

In recent years, the Federal Aviation Administration (FAA) has waded into uncharted legal territory by issuing Temporary Flight Restrictions (TFR) prohibiting citizens and news workers alike from flying drones above mass protests. Significantly, these restrictions do not apply to law enforcement agencies, effectively ceding airspace to local authorities and violating the First Amendment rights of journalists and activists. What's more, police act with impunity when firing on unmanned vehicles: a flagrant violation of FAA regulations. For critics, these no fly zones have "normalized a chilling precedent set by the FAA during the 2014 protests in Ferguson following the police killing of Mike Brown" (Koebler, 2016a). Against this backdrop, domestic law enforcement and border control agencies deploy drones to police local communities; monitor the movement of migrant populations; and conduct search and rescue missions (Wall & Monahan, 2011).

At the same time, news organizations, wildlife and resource management agencies, and whole swaths of the scientific community have adopted drone technology for all manner of surveillance activities (Marris, 2013; Qiu, 2014; Roug, 2014). Accordingly, this chapter employs a broad definition of surveillance to include "the multiple, complex socio-technological practices of intentional information gathering, especially when this information gathering is intended to support practices of control, organization, management or influence" (Barnard-Wills & Wells, 2012, p. 228). All of which is to suggest that as drones make their way from distant warzones to domestic airspace, these flying machines constitute new modes of surveillance: from controversial approaches to community policing, to novel innovations in drone journalism (Tremayne & Clark, 2013).

Media discourse surrounding drone surveillance expresses hope as well as fear—what James Carey describes as "opposite poles" of the sublime (1992, p. 207)—feelings long associated with the introduction of new technologies. On the one hand, photographers and journalists claim that cheap, easy to use flying cameras "democratize the sky" (Estrin, 2017). On the other hand, the sudden appearance of eyes in the sky raise privacy concerns for legal experts and civil liberties groups (Stanley, 2011). This chapter first examines media discourse surrounding the use of unmanned vehicles in domestic law enforcement. Specifically, I explore the relationship between the discourse of national security that legitimates the use of weaponized drones in America's war on terror, and the discursive strategies and practices employed by police departments and border control agencies to justify their use of surveillance drones. In addition, I consider objections to drone surveillance raised by activists, legal analysts and others.

Following this, I review arguments for and against the use of drones in electronic newsgathering. Examples gleaned from news organizations, as well as independent and citizen journalists, demonstrate how drone technology is changing journalistic practice—and reigniting longstanding debates over journalism ethics (Culver, 2014). The chapter concludes with an assessment of a far more sanguine discourse surrounding the application of drone technology for scientific, environmental, and humanitarian purposes. Throughout, I underscore the varied, contradictory, and frequently contentious discourse surrounding the role unmanned vehicles play in existing and emerging domestic surveillance regimes.

Border Security and Domestic Law Enforcement

The episode at Standing Rock reservation was not the first time North Dakota law enforcement encountered drone technology. According to a wire service report, in June 2011, Nelson County Sheriff Kelly Janke deployed a SWAT team, a bomb squad, and, most remarkably, an unarmed Predator B aircraft to track down three gunmen wanted for cattle rustling (Bennett, 2011). Using the surveillance plane's imaging and sensor technology to search a 3,000-acre spread, police located the men and apprehended them without incident: the first documented case of law enforcement using a Predator drone to arrest criminal suspects. As Fox News (2014) later put it: "What began as a wild west-style cattle-stealing case may have ushered in a brave new world of law enforcement officials using drones to gather evidence to put Americans behind bars."

The same wire report further indicates that Predator drones operating out of Grand Forks Air Force Base in North Dakota were used in dozens of police surveillance operations, courtesy of US Customs and Border Protection (CBP). What's more, the story revealed the Federal Bureau of Investigation (FBI) and the Drug Enforcement Agency (DEA) have also employed drones in domestic investigations. As Brian Bennett (2011) notes: "The previously unreported use of its [CBP's] drones to assist local, state and federal law enforcement has occurred without any public acknowledgement or debate." Herein we can discern a troubling parallel with the use of drones in US counterterrorism operations: a campaign shrouded in secrecy and with little public input or oversight. Equally important, the use of sophisticated spy planes for law enforcement is part and parcel of a broader trend that has expanded domestic surveillance authority since the attacks of September 11, 2001 (Bloss, 2009).

Unbeknownst to the public, the Predator B began making regularly scheduled flights along the US-Mexico border in September 2005.[1] By 2009, unmanned vehicles were patrolling the skies along the 49th parallel separating the United States and Canada. News and official discourse supporting the Predator's deployment along America's borders routinely invokes national security prerogatives, while downplaying the aircraft's lethal connotations. US border agents deployed the drones as part of an increasingly militarized border control regime—ostensibly designed to stem the flow of illegal immigrants and drug smugglers (Axtman, 2005). Echoing his call for a technological revolution in modern warfare, President George W. Bush likened the plane's role in border security to US counterterrorism efforts: "Slowly but surely, technology is being employed up and down the border, and that's a key part of our strategy" (Axtman). Notwithstanding the sublime appeal of superior technology securing America's borders, the prospect of unmanned vehicles, so closely associated with the war on terror, surveilling domestic airspace yields a decidedly ambivalent response.

On the one hand, Todd Faser, a spokesman for the US Border Patrol, enthusiastically supports the use of UAVs for border security: "Our primary mission is to prevent terrorists and their weapons from entering our country, and we need the right combination of personnel, technology, and infrastructure" (Axtman). Revealingly, Faser fails to mention either illegal immigration or the illicit drug trade; rather, he invokes the war on terror to justify the use of drones for domestic surveillance. On the other hand, T. J. Bonner, president of the National Border Patrol Council, questions the costs and benefits of drone technology for border security: "You can put a lot of boots on the ground for $14 million," he said (Axtman). Bonner advocates more efficient use of financial resources and takes a refreshingly non-technological approach to immigration policy: one that aims to restrict American employers from hiring undocumented workers in the first place. Underscoring Bonner's concerns, a subsequent investigation by the Department of Homeland Security (DHS) inspector general found that drones have been "dubious achievers" in border control efforts (Sledge, 2015).

A story published in Canada's *Globe and Mail* reveals much the same ambivalence toward UAVs patrolling America's northern border. The paper's national security correspondent, Paul Koring, opens his report with two observations: first, the porous US-Canada border could be exploited by jihadist fighters determined to enter the United States; second, the clandestine nature of the US drone program skews public understanding of these unmanned vehicles. Intended to demystify drone technology, and thereby legitimate its application along the Canadian border, Koring's report, datelined Grand Forks, is at once reassuring and distressing. John Priddy, who oversees Predator surveillance operations for Customs and Border Protection, told the newspaper: "We're here to protect the nation from bad people doing bad things." Invoking an era when the US and Canada worked together to confront the so-called Red menace, Priddy continues: "This is the equivalent of the Cold War in terms of a new type of vigilance" (Koring, 2011). Framing the new surveillance regime in these terms, Priddy invokes collective memory of the global struggle against Communism to justify the use of Predator drones operating in domestic airspace.

Further down in the report, Koring acknowledges the more troubling aspects of this unprecedented expansion of domestic surveillance: "It feels a lot like Big Brother is watching. The data—video, radar and thermal imagery—streams back to control rooms and can be watched live or stored for later analysis." In an otherwise reassuring story about the national security concerns underpinning drone surveillance of the US-Canadian border, the specter of an Orwellian police state engaged in endless war produces a nagging ambivalence. Unsurprisingly, Orwell's vision of an all powerful surveillance state informs much of the discourse surrounding drones operating in domestic airspace. Speaking with *National Geographic*, Jay Stanley, a senior policy analyst for the American Civil Liberties Union (ACLU), describes a "nightmare scenario" wherein domestic law enforcement adopts the US military's practice of *persistent surveillance* that enables "authorities combining drone video and cell phone tracking to build up databases of people's routine comings and goings—databases they can then mine for suspicious behavior" (Horgan, 2013a). The constant monitoring and analysis of everyday life not only undermines reasonable expectations of privacy; it opens the door for ever more intrusive forms of preemptive law enforcement.

Stanley's disturbing vision of robotic surveillance and predictive policing may be the stuff of dystopian fiction. All the same, law enforcement agencies across the country are using drones, and assorted military technologies, with greater regularity; thanks, in large measure, to "terrorism grants" issued by the Department of Homeland Security (Johnson & Shank, 2017). As a result, "from 2012 to 2014, the Federal Aviation Administration considered requests from at least 19 police and sheriff's departments, as well as National Guard units in nine states, to fly drones in domestic airspace"

(Gould-Wartofsky, 2015). Anticipating public resistance to the use of police surveillance drones, local law enforcement employ two distinct communication strategies. The first aims to alleviate privacy concerns with assurances of limited or "tactical" uses of unmanned vehicles. For instance, when the Los Angeles Police Department acquired two Draganflyer X6 drones in 2014—"gifts" from their counterparts in Seattle, who backed off plans to deploy the aircraft amid a chorus of civil liberty complaints—authorities insisted the planes would only be used under "narrow and prescribed" circumstances (Rubin, 2014). Similarly, New York City Police Commissioner Bill Bratton assured the City Council that the NYPD was only considering the use of unmanned planes. "I'm supportive of the concept of drones, not only for police but for public safety in general" (Moore, 2014). Thus, local law enforcement adopts a policy of strategic ambiguity to mitigate public apprehension.

A less forthcoming but increasingly common strategy is to field the technology without public notice. For example, the Hennepin County Sheriff's Office in Minnesota began flying a surveillance drone for search and rescue missions in May 2016; despite previous statements indicating the drone would not be operational for several months. More troubling, according to published reports: "Law enforcement officials are able to use the drone to collect evidence without first obtaining a warrant, and they are still failing to explain the full extent of the program" (Richards, 2016). Likewise, the San Diego Sheriff's Department one-year pilot program launched "quietly." "It didn't go before an elected body and community meetings weren't held, flying in the face of earlier advice from the local ACLU chapter" (Whitlock, 2016). Here again, the parallel between the US government's secretive targeted assassination program and the use of drones for domestic law enforcement is instructive. Indeed, only recently did the Department of Defense acknowledge it has been flying unmanned planes to conduct domestic surveillance missions since 2006 (McCauley, 2016). In the intervening years, a host of federal agencies have deployed drones in domestic airspace—with little public input or oversight (Friedersdorf, 2016). All of which is to suggest that in the absence of transparency in federal, state, and local government—let alone an informed and robust debate—fear, suspicion, and distrust permeate the public imagination.

Talk of weaponizing drones for domestic law enforcement further undermines public confidence and heightens socio-political anxieties: a troubling state of affairs for a democratic society, but altogether ripe conditions for authoritarian rule. Once again, North Dakota made headlines when the state legislature authorized the use of police drones armed with "non-lethal" weapons: another first for US law enforcement (Ferenstein, 2015). According to the bill's author, Representative Rick Becker, the original legislation called for banning weaponized drones and required police to obtain a search warrant before conducting drone surveillance. Backed by lobbyists representing local law enforcement, state representatives amended the bill to allow police to equip drones with Tasers and rubber bullets in exchange for the provision on search warrants (Glawe, 2015).

The prospect of local law enforcement deploying armed UAVs prompted Christof Heyns, the United Nations special rapporteur on extrajudicial, summary or arbitrary executions, to declare: "An armed drone, controlled by a human from a distance, can hardly do what police officers are supposed to do—use minimum force required by the circumstances" (Cohn, 2015). While corporate news organizations provided extensive coverage of North Dakota's landmark legislation, few outlets noted the role industry lobbyists played in drafting the bill, let alone the legal and humanitarian concerns expressed by Heyns. In rather duplicitous fashion, then, North Dakota ushered in the era of weaponized drones in domestic law enforcement: an idea that has since gained currency after the Dallas Police Department used a

Figure 5.1: Pedestrians photograph Essam Attia's satirical NYPD drone advertisement. *Photo credit: Jason Eppink. Source: Flickr.*

robot armed with explosives to kill a gunman who shot and killed five police officers in July, 2016 (Sanchez, 2016). For Ryan Calo, an expert in cyber and robotic law at the University of Washington School of Law, the implications are clear: "It is essentially a jury-rigged version of a drone strike" (Koebler, 2016b).

In an instance of life imitating art, New York City graphic artist Essam Attia anticipated many of these developments—albeit in hyperbolic fashion—in September 2012. Alarmed by the expansive surveillance authority the NYPD assumed in recent years, and ongoing efforts by DHS to outfit local police departments with surveillance drones, Attia and his accomplices placed a series of spoof public service advertisements (PSAs) on telephone kiosks around Manhattan. Noteworthy for its daring, its gallows humor, and its intertextual dexterity, the faux campaign satirizes police surveillance programs. One poster features a Big Brother figure, flanked by surveillance cameras and unmanned planes, looming over a group of terrified citizens. Set against a blue background, complete with seemingly authentic NYPD and Advertising Council logos, the tagline reads: "Always Watching." Another PSA parody depicts a family fleeing for their lives as a Predator aircraft fires a missile. The tagline reads: "Drones: Protection When You Least Expect It."

A veteran who served two tours in Iraq, Attia explained that his provocation aimed to start a "conversation" over the NYPD's surveillance programs (Harvey & Ismail, 2012). Like the sporadic debate over America's targeted killing program discussed in Chapters 2 and 3, this vital discussion has been thwarted by government secrecy on the one hand, and indifferent news reporting on the other. Attia's satirical intervention sought to generate public awareness and debate over domestic surveillance (Figure 5.1). When police finally caught up with Attia, some three months after the posters first appeared, an embarrassed NYPD threw the book at him (Opelka, 2012). But in a victory for freedom of expression, authorities eventually dropped the charges. Still, we should not underestimate the significance of Attia's hoax. Not only did the satirical campaign articulate widespread apprehension over a nascent American police state. It vividly identified and forcefully challenged the lack of transparency and public oversight of an emerging surveillance regime. Herein lies the political value of ironic media activism—a subject taken up at length in Chapters 10 and 11. For now, we turn to consider the uses (and potential abuses) of drones in electronic newsgathering, citizen journalism, and investigative reporting.

Drone Journalism

Not to be mistaken with news stories about unmanned aerial vehicles, drone journalism refers to the use of flying cameras to fulfill a basic function of mass media: surveillance of the environment (Lasswell, 1948; Wright, 1960). Of course, aerial photography acquired with helicopters and fixed-wing aircraft has been a mainstay of the news industry for decades. But like most innovations, the less proximate antecedents of drone journalism are frequently overlooked or forgotten. In the first decade of the last century, George R. Lawrence, an iconoclastic photographer from Chicago, designed what he called a captive airship. Lawrence mounted his bulky camera to a stack of seventeen kites capable of reaching heights up to 2000 feet. This "proto-drone" produced stunning airborne panoramas, including unrivaled documentary images of the 1906 San Francisco earthquake (Dawidowicz, 2013). A pioneer of aerial photography, Lawrence's work anticipates the applications, as well as the aesthetics, of contemporary drone journalism.

Like Lawrence before them, advocates of drone journalism leverage the technology's unique ability to produce (and relay) images from locations that are too dangerous or otherwise inaccessible for photographers working on the ground. Observing the risks news workers take when reporting from battlefields and other conflict areas, Matt Schroyer, a data journalist and early advocate of drone journalism, contends:

> If reporters are able to launch drones in the midst of an uprising, and capture with an 'eye in the sky' oppression or crimes against humanity, potentially to a live, international television or internet audience, it would revolutionize the way events abroad are reported and realized. (Schroyer, 2011)

Like troops operating military drones, unmanned aircraft keep journalists out of harm's way. Moreover, the technology enables reporters to circumvent media blackout zones imposed by the private interests and the state. Drone journalism also has the potential to foster collaboration between news workers, scientists and others for data-driven stories and investigative reports. For example, drones fitted with cameras and sensors can assess air and water quality "or take stock of contaminants and pollutants near an industrial plant" (Schroyer). Drones offer similar advantages for covering natural disasters and breaking news events (Tremayne & Clark, 2013). In this way, drones dramatically enhance journalism's surveillance function, with clear and compelling public interest implications. Professor Matt Waite, founding director of the

Drone Journalism Lab at the University of Nebraska-Lincoln, sums up the advantages of the new technology this way: "I think drones—small, cheap, easy-to-use vehicles that can fit in a small bag and carried into the field by a reporter—offer a major opportunity to improve certain kinds of reporting" (Martinelli, 2011). Of course, along with these opportunities come formidable challenges (Goldberg, Corcoran, & Picard, 2013).

Writing for WorldViews, an online forum hosted by foreign correspondents for the Washington *Post*, reporter Melissa Bell considers the immediate and future prospects of drone journalism. Invoking a familiar trope, she begins: "This January, the FAA will be proposing new rules on the use of drones in American airspace—a possibility some see as positively Orwellian, but others, including some journalists, see as an opportunity" (Bell, 2011). Summarizing the positions of drone journalism enthusiasts Matt Waite and Matthew Schroyer, Bell further notes the economic incentive for news organizations to acquire the technology: "For shrinking newsrooms [sic] staffs, drones that cost around $40,000 sound a bit more budget friendly than helicopters that cost in the millions." Conversely, Bell alerts readers to the potential downside of drone journalism, such as privacy and ethical abuses that could compromise journalistic integrity. "The technology could easily be taken advantage of by celebrity trackers. It could also mean journalists could be kept under surveillance as well" (Bell). In short, the column offers readers a breezy and relatively upbeat assessment of the limits and possibilities of drone journalism—just the thing for the *Post's* interactive news portal. Nevertheless, a sampling of reader comments reveals a far more jaundiced perspective on the matter.[2]

> Haven't seen real "journalism" in America for decades, only War Cheerleading, torture justifying, death and celebrity poison.
>
> We already have plenty of drone journalism, particularly in broadcasting. Just watch or listen and experience the same stories repeated over and over and over. They just drone on and on. Why don't they spend the money for some real investigative reporting?

These comments register deep frustration with the state of contemporary American journalism and healthy skepticism toward a technological fix to the problems of the Fourth Estate. The first laments a dearth of critical reporting, especially in the conduct of America's war on terror, and underscores the crucial role news media play in legitimating US militarism—a subject we consider at length in the following chapter. The second revels in a bit of word play to level an equally valid critique of American news culture. In unmistakable reference to rolling news channels, the reader challenges an economic model of journalism predicated on entertainment, spectacle, and diversion, rather than the public interest. Media criticism aside, readers also expressed grave misgivings over the relationship between surveillance technologies and creeping authoritarianism manifest by these flying, robotic cameras.

> Better close your curtains, these things have very intrusive technology. Only a matter of time before they are armed and aimed at dissenters here in the good ol USA.

> Drone journalism is also another form of public intrusion, spying, and aerial surveillance.

As we noted in Chapter 4, the discourse of domestication aims to alleviate concerns over the privacy and public safety implications of drone technology. But as these comments demonstrate, the fear associated with the military and surveillance connotations of these flying machines continue to shape and inform public opinion (Scott, 2015). It remains to be seen whether or not the new technology will, in fact, enhance and improve press performance and practice. In the meantime, we can detect a few trends: some promising, others more troubling.

News Corporation's now defunct *The Daily* was among the first media organizations to use drones for electronic newsgathering. In May 2011, reporters using a MicroDrone MD4-1000 acquired dramatic aerial footage of tornado damage across Alabama. The material was used for a new feature called "The Daily Drone" (Gell, 2011). The following month, "The Daily Drone" documented massive flooding in Minot, North Dakota. The possibilities for using unmanned vehicles to enhance news coverage of storm devastated communities quickly became apparent. So too did potential legal violations involved in obtaining the footage. At the time, the FAA prohibited commercial operation of drones; as a result, the agency opened an investigation into *The Daily's* activities (Hill, 2011). As noted in the previous chapter, the FAA issued new rules governing commercial drone operations in August 2016. The new regulations make it easier for journalists to obtain a remote pilot certificate, rather than a traditional pilot's license: an expensive and time consuming process that deterred veteran reporters and aspiring journalists alike from using drones for newsgathering (Lee, 2015). In the intervening years, however, a slow moving regulatory process, further complicated by debates over journalism's privileged status in the Constitution versus its commercial orientation, stymied the development of drone journalism in the United States (Mortimer, 2015). But as FAA prohibitions did not apply to hobbyists, the regulatory vacuum opened up "DIY drone space" for activists, citizen journalists and others to shape the emerging field of drone journalism (Cassimally, 2013).

For instance, in late 2011, as the NYPD launched a crackdown on the Occupy Wall Street encampment in Zuccotti Park, videographer Tim Pool used his makeshift "Occucopter" to document police activity (Martin, 2011). During the evacuation, police barred journalists from the area—declaring it a "frozen zone"—ostensibly out of safety concerns. Determined to circumvent a news blackout surrounding the forcible eviction of nonviolent protestors, Pool launched the device to live stream and document the melee. Speaking with *The Guardian*, Pool argued that his flying camera "provides a clever loophole around certain restrictions such as when the police block press from taking shots of an incident" (Sharkey & Knuckey, 2011). Pool's innovative approach to counter surveillance anticipates similar practices, from the 2014 Black Lives Matter protests in Ferguson, Missouri, to the 2016 NoDAPL resistance at Standing Rock. But not, as we have seen, without pushback from law enforcement. Indeed, the FAA's acquiescence to increasingly frequent police requests for Temporary Flight Restrictions will likely have a chilling effect on professional and citizen journalists alike:

> As a new era of civil resistance dawns under the Trump administration, at the Standing Rock site and in anti-Trump demonstrations across the country, drone experts say police and government have made it unnecessarily difficult—sometimes impossible—for civilians to deploy drones at large protests. (Kopstein)

Police are not the only subject of drone-enabled, citizen surveillance regimes. In Oregon, the Eastmoreland Neighborhood Association uses a flying camera to keep an eye on the Union Pacific rail yard in the midst of a legal battle between residents and the railroad over complaints of noise and air pollution (Binder, 2014). And in rural Pennsylvania, Showing Animals Respect and Kindness (SHARK), flies a drone to document illegal pigeon shoots and other instances of animal cruelty, inspiring likeminded activist groups to do the same (Russell, 2013). One of the most celebrated instances of these increasingly common surveillance practices led to a criminal investigation of a meat packing plant outside of Dallas, Texas. Using an Exposure airframe vehicle, an unnamed hobbyist inadvertently documented environmental pollution on an industrial scale. Speaking with *sUAS News*, the pilot said: "I was looking at images after the flight that showed a blood red creek and was thinking, could this really be what I think it is? Can you really do that, surely not?" (Mortimer, 2012). Opponents of so-called ag-gag bills prohibiting journalists from investigating factory farms have since launched a crowdfunding campaign, "Drone on the Farm: An Aerial Expose," to circumvent the law and document potential criminal activity (Zara, 2014).

None of which is to suggest that any of these practices are free from ethical considerations. The potential for abuse of drone technologies in pursuit of a news story is evident. Paparazzi using drones—so-called dronearazzi—are high on the list of likely offenders. So too, breaking news reporters must exercise caution when weighing the potential news value of drone footage with longstanding concerns over journalism ethics, privacy rights, public safety, and conflicts of interest (Culver). For instance, in Hartford, Connecticut, a stringer for a local TV station used a UAV equipped with a video camera to capture graphic images of a fatal automobile accident "in which the victim's body was left hanging out of a mangled car" (Melia, 2014). Forsaking good taste and common decency for sensational footage, reporters who abuse the new technology undermine the promise drone journalism holds for surveilling the environment in the public interest.

Hence the establishment, under the direction of the Professional Society of Drone Journalists, of a code of conduct for the emerging field. Underscoring the need to balance news values with public safety and the law, privacy rights and professional responsibilities, the code of ethics does not set out to reinvent the wheel. Rather, it provides a framework to guide drone journalists in pursuit of their craft. The document reads, in part, "A code of ethics for drone journalists should be viewed as a layer of additional ethical considerations atop the traditional professional and ethical expectations of a journalist in the 21st century" (PSDJ, n.d.). All of which underscores an important point: clear ethical guidelines and sound judgment must determine how this, or any other technological innovation, is put to use. Not the other way around.

Drones for Good

From killing by remote control to opening up new vistas for photojournalists, drones change how we relate to distant others, how we tell our stories, and, quite literally, how we see the world. Amid news reports of US drone strikes, warrantless police searches, and all manner of airborne criminality—from drug trafficking and spying on the neighbors to shooting down your neighbor's quadcopter—the hazards of drone technology frequently overshadow its benefits. More to the point, media discourse surrounding drones frequently arouses fear and loathing in the public mind (Horgan, 2013b). Nevertheless, as promoters, enthusiasts, and a growing number of journalists remind us: drones can be used for good. Conservationists leverage the unique aerial perspectives drones provide to monitor endangered species and fragile ecosystems. Climate scientists launch drones to study extreme weather conditions and measure sea ice at the poles. And humanitarian organizations fly drones to coordinate disaster relief efforts and document human rights abuses. In short, media accounts of what I'm calling, for lack of a better phrase, "benign regimes of surveillance," play an increasingly prominent and potentially decisive role in demystifying drones for an otherwise wary public. Briefly, then, I want to consider how news accounts of benevolent drones contribute to a discourse of domestication.

In 2014, the Nature Conservancy used a consumer-grade DJI Phantom quadcopter to monitor and record the population of the sandhill crane, a threatened migratory bird and frequent wintertime visitor to the California Delta. Three years earlier, the group purchased a piece of land that has come to serve as a sanctuary and a living laboratory to study and protect the species. In years past, volunteers struggled to come up with an accurate count of birds visiting the refuge in winter months. This time around, a single overhead drone, outfitted with daytime and infrared cameras for nighttime photography, produced detailed images yielding a more precise headcount. With clearer regulations governing the use of unmanned vehicles for environmental research, conservationists see enormous potential for "crane counting with a drone" (May, 2014).

Meanwhile, marine biologists at the Woods Hole Oceanographic Institute fly "whale breath-catching drones" to conduct microbial analysis of the creatures. Like a growing number of conservationists, researchers off the Massachusetts coast have realized the potential drones offer for studying animal health and behavior. "The availability of reliable, off-the-shelf UAVs has made standard aerial surveillance far easier than it used to be, but drones don't come out of the box ready for the kinds of projects conservationists throw at them" (Baraniuk, 2016). Nevertheless, with a bit of DIY ingenuity, scientists and nature enthusiasts alike can conduct research in a safe, timely, environmentally friendly fashion—all at a fraction of the cost of more conventional research methods (Baggaley, 2017). Equally important, surveillance drones patrolling the skies over the habitat of endangered species provide an important line of defense against animal poaching. Rather than taking or threatening lives, drone technology serves to protect wildlife against human avarice and environmental degradation.

In Oklahoma's notorious "tornado alley" researchers launch specially designed UAVs to study and better predict severe weather conditions. Thanks to a $6 million grant from the National Science Foundation, meteorologists at four universities across the state fly weather sensing drones remarkably close to storms, enabling scientists to collect data on atmospheric pressure, wind speed and direction, thereby improving the accuracy and timeliness of severe weather forecasts. Alyssa Avery, a graduate student at Oklahoma State, designed such an aircraft to "fly around severe storms and collect as much data as we can so we can lengthen that warning time and make it safer for everyone living in tornado areas" (Fee, 2016). Meteorologists and climatologists have used satellites, manned aircraft, and all manner of UAVs for years. However, recent developments in drone technology represent a dramatic improvement in data collection for weather forecasts and climate research. According to David Johnson, an environmental scientist at Duke University, the rapid turn around time alone makes drones an incredibly useful research tool, producing "workable data in four hours, versus maybe four days if we're using a plane" (O'Connor, 2016). Capitalizing on the growing recreational drone market, climate scientists have also turned to crowdsourcing data collection. Working in close cooperation with a small army of "citizen scientists," researchers in California developed a "distributed surveillance strategy" to study rising sea level during a recent El-Nino event (Vlasits, 2017). News stories such as these highlighting the potential for drones to protect the environment and enhance public safety during extreme weather conditions go a long way toward normalizing the use of drones in domestic airspace.

Finally, in sharp contrast to the dominant order of discourse surrounding drones in America's targeted killing program, humanitarian organizations contend that we can use this technology to save lives. And indeed, it has. Writing in the New York *Times*, Andrew Stobo Sniderman and Mark Hanis, co-founders of the Genocide Intervention Network, make a compelling argument that the same technologies that allow conservationists to monitor and document illegal whaling operations on the high seas ought to be used to surveil authoritarian regimes committing crimes against humanity. Unlike expensive and sometimes difficult to read satellite images of conflict zones, drones offer activists an economical and easy-to-use surveillance device for documenting human rights abuses. Echoing the arguments of drone journalism enthusiasts, humanitarian groups suggest this footage could be broadcast or live streamed to an international audience and sway public opinion. "We could record the oppression in Syria with unprecedented precision and scope. The better the evidence, the clearer the crimes, the higher the likelihood that the world would become as outraged as it should be" (Sniderman & Hanis, 2012).

Leaving aside for a moment the role covert operations and foreign military interventions have played in intensifying and prolonging the Syrian conflict, surveillance drones have been instrumental in locating and monitoring war refugees making the perilous journeys to Europe across the Mediterranean Sea. According to a report in *sUAS News*, the humanitarian NGO Migrant Offshore Aid Station (MOAS) has saved over 8000 lives, thanks, in large part, to the surveillance capabilities of the Camcopter S-100, an unmanned helicopter.

> The S-100 proved to be a huge asset during the operation as it can locate refugee boats by day and night, even in rough sea conditions and at long distance. The camera of the unmanned helicopter delivers daylight and infrared video in real time to the MOAS team. (Press, 2015)

There is, of course, a cruel irony to this. Weaponized drones—the embodiment of the new kind of war George W. Bush advocated; Barack Obama's weapon of choice in US counterterrorism operations—help fuel the refugee crisis across the Greater Middle East. A crisis which, in all likelihood, will grow worse under the Trump administration's expanded use of airstrikes across the region (Kheel, 2017). According to published reports, within days of taking office, President Trump eased restrictions on US forces designed to minimize civilian casualties (Savage & Schmitt, 2017). As a result, civilian deaths in airstrikes conducted by both manned and unmanned vehicles escalated dramatically during Trump's first 100 days in office. All of which begs the question: Why is the American public, at once fearful and fascinated by the prospect of drones flying in domestic airspace, so complacent about the US drone program? The answer lies, in large measure, with the character and conduct of domestic news coverage of the drone wars.

Notes

1. From October 29 through November 12, 2003, the Bureau of Immigration and Customs Enforcement (ICE) ran a classified pilot program using unarmed Predators, Operation Safeguard, along the US-Mexico border. See "ICE deems drone program 'isn't news.'"

Muckrock, https://www.muckrock.com/news/archives/2015/mar/16/ice-deems-drone-program-isnt-newsworthy/.

2. See "Comments," https://www.washingtonpost.com/blogs/blogpost/post/drone-journalism-the-idea-could-fly-in-the-ussoon/2011/12/04/gIQAhYfXSO_blog.html#comments.

References

Axtman, K. (2005, December 7). US border patrol's newest tool: a drone. *Christian Science Monitor*. Retrieved from http://www.csmonitor.com/2005/1207/p02s01-usmi.html

Baggaley, K. (2017, February 10). Drones are setting their sights on wildlife. *Popular Science*. Retrieved from http://www.popsci.com/drones-wildlife-biology-animal-research

Baraniuk, C. (2016, November 13). Modified drones are keeping an eye on the world's wildlife. *Quartz*. Retrieved from https://qz.com/829983/modified-drones-are-keeping-an-eye-on-the-worlds-wildlife/

Barnard-Wills, D., & Wells, H. (2012). Surveillance, technology and the everyday. *Criminology & Criminal Justice*, *12*(3), 227–237.

Bell, M. (2011, December 4). Drone journalism? The idea could fly in the US. *Washington Post*. Retrieved from https://www.washingtonpost.com/blogs/blogpost/post/drone-journalism-the-idea-could-fly-in-the-ussoon/2011/12/04/gIQAhYfXSO_blog.html

Bennett, B. (2011, December 10). Police employ Predator drone spy planes on home front. *Los Angeles Times*. Retrieved from http://articles.latimes.com/2011/dec/10/nation/la-na-drone-arrest-20111211

Binder, M. (2014, June 9). SE neighborhood buys a drone, plans to spy on Union Pacific, developers. *OregonLive.com*. Retrieved from http://www.oregonlive.com/portland/index.ssf/2014/06/se_neighborhood_buys_a_drone_p.html

Bloss, W. P. (2009). Transforming US police surveillance in a new privacy paradigm. *Police Practice and Research*, *10*(3), 225–238.

Carey, J. W. (1992). *Communication as culture: Essays on media and society*. New York, NY: Routledge.

Cassimally, K. A. (2013, February 28). Brace yourselves, drone journalism is coming. *Scientific American*. Retrieved from https://blogs.scientificamerican.com/incubator/brace-yourselves-drone-journalism-is-coming/

CBS/AP. (2016, October 24). Drone shot down, dozens arrested in weekend Dakota Access pipeline protests. *CBS News*. Retrieved from http://www.cbsnews.com/news/drone-shot-down-dozens-arrested-dakota-access-pipeline-protests/

Cohn, M. (2015, September 1). One day soon, that drone overhead may be pointing a taser at you. *Truthdig*. Retrieved from http://www.truthdig.com/report/item/cohn_drones_watching_you_20150901

Culver, K. B. (2014). From battlefield to newsroom: Ethical implications of drone technology in journalism. *Journal of Mass Media Ethics*, *29*, 52–64.

Dawidowicz, J. (2013, July 1). Century-old "proto-drone" shot gorgeous aerial panoramas of Brooklyn and San Francisco. *Animal*. Retrieved from http://animalnewyork.com/2013/century-old-proto-drone-shot-gorgeous-aerial-panoramas-of-brooklyn-and-san-fransisco/

Ellis, R., & Grinberg, E. (2016, October 24). Drone shot, road reopens at Dakota Access Pipeline protest. *CNN*. Retrieved from http://www.cnn.com/2016/10/22/us/dakota-access-pipeline-arrests/

Estrin, J. (2017, March 16). Democratizing the sky: Drones in visual journalism. *New York Times*. Retrieved from https://lens.blogs.nytimes.com/2017/03/16/democratizing-the-sky-drones-in-visual-journalism/

Fee, S. (2016, February 21). In Tornado Alley, using drones to pinpoint severe weather. *PBS NewsHour*. Retrieved from http://www.pbs.org/newshour/bb/in-tornado-alley-using-drones-to-pinpoint-severe-weather/

Ferenstein, G. (2015, August 26). Weaponized drones for law enforcement now legal in North Dakota. *Forbes*. Retrieved from https://www.forbes.com/sites/gregoryferenstein/2015/08/26/weaponized-drones-now-legal-inside-the-u-s-lawmaker-says-crimefighting-will-become-a-video-game/#67b8592f50c1

Fox News. (2014, January 28). Predator drone helps convict North Dakota farmer in first case of its kind. *Fox News.com*. Retrieved from http://www.foxnews.com/us/2014/01/28/first-american-gets-prison-with-assistance-predator-drone.html

Friedersdorf, C. (2016, March 10). The rapid rise of federal surveillance drones over America. *The Atlantic*. Retrieved from https://www.theatlantic.com/politics/archive/2016/03/the-rapid-rise-of-federal-surveillance-drones-over-america/473136/

Gell, A. (2011, May 2). "The Daily" finally unleashes those drone choppers! *Observer*. Retrieved from http://observer.com/2011/05/the-daily-finally-unleashes-those-drone-choppers/

Glawe, J. (2015, August 26). First state legalizes Taser drones for cops, thanks to a lobbyist. *The Daily Beast*. Retrieved from http://www.thedailybeast.com/articles/2015/08/26/first-state-legalizes-armed-drones-for-cops-thanks-to-a-lobbyist.html

Goldberg, D., Corcoran, M., & Picard, R. G. (2013). *Remotely piloted aircraft systems & journalism: Opportunities and challenges of drones in news gathering*. Reuters Institute for the Study of Journalism, University of Oxford.

Gould-Wartofsky, M. (2015). A 5-step guide to policing a protest movement in post-9/11 America. *Moyers & Company*. Retrieved from http://billmoyers.com/2015/05/07/wars-come-home/

Harvey, M., & Ismail, A. (2012, September 24). Wanted "drone" poster artist discusses how he punked the NYPD. *Animal*. Retrieved from http://animalnewyork.com/2012/wanted-drone-poster-artist-discusses-how-he-punked-the-nypd/

Herzog, C. (2016, December 1). #NoDAPL protest effectively wields drones and social media. *WhiteHat Magazine*. Retrieved from http://whitehatmag.com/2016/12/01/drones-social-media-standing-rock/

Hill, K. (2011, August 2). FAA looks into News Corp's Daily Drone, raising questions about who gets to fly drones in the US. *Forbes*. Retrieved from https://www.forbes.com/sites/kashmirhill/2011/08/02/faa-looks-into-news-corps-daily-drone-raising-questions-about-who-gets-to-fly-drones-in-the-u-s/#33274c6633b1

Horgan, J. (2013a, March). Unmanned flight: The drones come home. *National Geographic*. Retrieved from http://ngm.nationalgeographic.com/2013/03/unmanned-flight/horgan-text

Horgan, J. (2013b, February 16). Why drones should make you afraid. Very afraid. *Scientific American*. Retrieved from https://blogs.scientificamerican.com/cross-check/why-drones-should-make-you-afraid-very-afraid/

Johnson, H., & Shank, M. (2017, March 17). Dumping military kit on cops makes main street a war zone. *Newsweek*. Retrieved from http://www.newsweek.com/dumping-military-kit-cops-makes-main-street-war-zone-568889

Kheel, R. (2017, March 27). Reports of civilian deaths raise questions about Trump's airstrike policy. *The Hill*. Retrieved from http://thehill.com/policy/defense/326020-reports-of-civilian-deaths-raise-questions-about-trumps-airstrike-policy

Koebler, J. (2016a, November 30). The government is using a no fly zone to suppress journalism at Standing Rock. *Motherboard*. Retrieved from https://motherboard.vice.com/en_us/article/the-government-is-using-a-no-fly-zone-to-suppress-journalism-at-standing-rock

Koebler, J. (2016b, July 2016). The legal and ethical ramifications of letting police kill suspects with robots. *Motherboard*. Retrieved from https://motherboard.vice.com/en_us/article/dallas-shooting-bomb-robot-legal-analysis

Kopstein, J. (2017, January 27). Police are making it impossible to use drones to document protests. *Vocativ*. Retrieved from http://www.vocativ.com/396662/police-drone-journalists-protests/

Koring, W. (2011, September 30). Canada-US border-patrolling aircraft hunt "bad people doing bad things." *Globe and Mail*. Retrieved from http://www.theglobeandmail.com/news/world/canada-us-border-patrolling-aircraft-hunt-bad-people-doing-bad-things/article556856/

Lasswell, H. (1948). *The structure and function of communication and society: The communication of ideas*. New York, NY: Institute for Religious and Social Studies.

Lee, D. (2015, August 28). Why a drone journalism educator is getting his pilot's license. *Columbia Journalism Review*. Retrieved from http://www.cjr.org/united_states_project/matt_waite_drone_journalism_lab.php

Listening Post. (2016, December 11). Dakota Access Pipeline: Behind the "media blackout." *Al Jazeera English*. Retrieved from http://www.aljazeera.com/programmes/listeningpost/2016/12/dakota-access-pipeline-media-blackout-161210071122040.html

Marris, E. (2013, June 12). Drones in science: Fly, and bring me data. *Nature*. Retrieved from http://www.nature.com/news/drones-in-science-fly-and-bring-me-data-1.13161

Martin, A. (2011, December 7). Occupy Wall Street has a drone: The occucopter. *The Atlantic*. Retrieved from https://www.theatlantic.com/national/archive/2011/12/occupy-wall-street-has-drone-occucopter/324966/

Martinelli, N. (2011, December 19). Five things you need to know about drone journalism. *IJNet*. Retrieved from http://ijnet.org/en/blog/five-things-you-need-know-about-drone-journalism

May, P. (2014, December 1). Drone helps California conservation group with the threatened sandhill crane. *The Mercury News*. Retrieved from http://www.mercurynews.com/2014/12/01/drone-helps-california-conservation-group-with-the-threatened-sandhill-crane/

McCauley, L. (2016, March 9). War zone tactics come home as Pentagon admits domestic spy drone use. *Common Dreams*. Retrieved from http://www.commondreams.org/news/2016/03/09/war-zone-tactics-come-home-pentagon-admits-domestic-spy-drone-use

Melia, M. (2014, February 11). Drone use highlights questions for journalists. *USA Today*. Retrieved from http://www.usatoday.com/story/tech/2014/02/11/us-drones-faa-media/5390155/

Moore, T. (2014, May 20). NYPD considering using drones to fight crime. *New York Daily News*. Retrieved from http://www.nydailynews.com/new-york/nyc-crime/nypd-drones-fight-crime-article-1.1799980

Mortimer, G. (2012, January 23). Dallas meat packing plant investigated after drone images reveal pollution. *sUAS News*. Retrieved from https://www.suasnews.com/2012/01/dallas-meat-packing-plant-investigated-after-drone-images-reveal-pollution/

Mortimer, G. (2015, May 7). Citizen drone journalists can be paid, others not so much. *sUAS News*. Retrieved from https://www.suasnews.com/2015/05/citizen-drone-journalists-can-be-paid-others-not-so-much/

O'Connor, M. C. (2016, April 19). Why scientists are banking on drones for tracking coastal climate research. *Pacific Standard*. Retrieved from https://psmag.com/why-scientists-are-banking-on-drones-for-tracking-coastal-climate-research-ce56a3280ae5

Opelka, M. (2012, December 9). New York cops finally nab artist who posted embarrassing "NYPD Drone" posters. *The Blaze*. Retrieved from http://www.theblaze.com/news/2012/12/09/new-york-cops-finally-nab-artist-who-posted-embarrassing-nypd-drone-posters/

Press. (2015, October 2). CAMCOPTER S-100 assisted in the rescue of over 8,800 refugees in the Mediterranean. *sUAS News*. Retrieved from https://www.suasnews.com/2015/10/camcopter-s-100-assisted-in-the-rescue-of-over-8800-refugees-in-the-mediterranean/

PSDJ. (n.d.). Code of Ethics. *Professional Society of Drone Journalists*. Retrieved from http://www.dronejournalism.org/code-of-ethics/

Qiu, L. (2014, November 15). Watch: Can drones help save wildlife around the world? *National Geographic*. Retrieved from http://news.nationalgeographic.com/news/2014/11/141114-drones-wildlife-poaching-animals-conservation/

Richards, S. (2016, August 24). Drones: Law enforcement's secret eye in the sky. *Think Progress*. Retrieved from https://thinkprogress.org/drones-law-enforcements-secret-eye-in-the-sky-9eda68fb8bc#.v2e01fb96

Roug, L. (2014, May 1). Eye in the sky. *Columbia Journalism Review*. Retrieved from http://archives.cjr.org/cover_story/eye_in_the_sky.php

Rubin, J. (2014, May 30). LAPD adds drones to arsenal, says they'll be used sparingly. *Los Angeles Times*. Retrieved from http://www.latimes.com/local/lanow/la-me-ln-lapd-adds-drones-to-arsenal-20140530-story.html#

Russell, L. (2013, April 12). PETA eyes drones to watch hunters, farmers. *CNN.com*. Retrieved from http://www.cnn.com/2013/04/11/us/animal-rights-drones/

Sanchez, C. (2016, October 20). Police departments are thinking of using drones armed with stun guns. *Business Insider*. Retrieved from http://www.businessinsider.com/police-drones-armed-with-stun-guns-taser-2016-10

Sandhu, A., & Haggerty, K. D. (2017). Policing on camera. *Theoretical Criminology*, *21*(1), 78–95.

Savage, C., & Schmitt, E. (2017, March 30). Trump eases combat rules in Somalia intended to protect civilians. *New York Times*. Retrieved from https://www.nytimes.com/2017/03/30/

world/africa/trump-is-said-to-ease-combat-rules-in-somalia-designed-to-protect-civilians.html?_r=0

Schroyer, M. (2011, May 26). A radical new mission for drones: Helping journalists find truth. *The Mental Munition Factory*. Retrieved from http://www.mentalmunition.com/2011/05/radical-new-mission-for-drones-helping.html

Scott, A. (2015, February 5). Americans OK with police drones—private ownership, not so much: Poll. *Reuters*. Retrieved from http://www.reuters.com/article/us-usa-drones-poll-idUSKBN0L91EE20150205

Sharkey, N., & Knuckey, S. (2011, December 21). Occupy Wall Street's "occucopter"—who's watching whom? *The Guardian*. Retrieved from https://www.theguardian.com/commentisfree/cifamerica/2011/dec/21/occupy-wall-street-occucopter-tim-pool

Sledge, M. (2015, January 6). Homeland Security drones get "dubious" results at 5 times advertised costs. *The Huffington Post*. Retrieved from http://www.huffingtonpost.com/2015/01/06/homeland-security-drones_n_6424370.html

Sniderman, A. S., & Hanis, M. (2012, January 30). Drones for human rights. *New York Times*. Retrieved from http://www.nytimes.com/2012/01/31/opinion/drones-for-human-rights.html

Stanley, J. (2011, December 15). New eyes in the sky: Protecting privacy from domestic drone surveillance. *American Civil Liberties Union*. Retrieved from https://www.aclu.org/blog/new-eyes-sky-protecting-privacy-domestic-drone-surveillance

Tremayne, M., & Clark, A. (2013). New perspectives from the sky. *Digital Journalism*. doi:10.1080/21670811.2013.805039

Vlasits, A. (2017, February 17). Drones are turning civilians into an Air Force of citizen scientists. *Wired*. Retrieved from https://www.wired.com/2017/02/drones-turning-civilians-air-force-citizen-scientists/

Wall, T., & Monahan, T. (2011). Surveillance and violence from afar: The politics of drones and liminal security-scapes. *Theoretical Criminology, 15*(3), 239–254.

Whitlock, J. (2016, December 15). The sheriff's department quietly launched a drone program with no public input. *Voice of San Diego*. Retrieved from http://www.voiceofsandiego.org/topics/public-safety/sheriffs-department-quietly-launched-drone-program-no-public-input/

Wright, W. R. (1960). Functional analysis and mass communication. *Public Opinion Quarterly, 24*, 610–613.

Zara, C. (2014, June 12). Fighting ag-gag laws with drones? Journalist eyes the skies for factory-farm investigations. *IBT*. Retrieved from http://www.ibtimes.com/fighting-ag-gag-laws-drones-journalist-eyes-skies-factory-farm-investigations-1599912

Zilber, A. (2016, October 24). North Dakota sheriffs shoot down drone that approached a helicopter in a "threatening manner." *Daily Mail*. Retrieved from http://www.dailymail.co.uk/news/article-3868292/North-Dakota-sheriffs-shoot-drone-approached-helicopter-threatening-manner-dozens-arrested-protests-Dakota-Access-Pipeline-past-weekend.html

· 6 ·

REPORTING THE DRONE WARS

Virtually every time the US fires a missile from a drone and ends the lives of Muslims, American media outlets dutifully trumpet in headlines that the dead were "militants"—even though those media outlets literally do not have the slightest idea of who was actually killed.

—Glenn Greenwald

This chapter considers the relationship between reporting in US corporate media and public knowledge and attitudes toward the drone wars. Drawn from democratic theories of the press, I employ the following criteria for evaluating this coverage. First, journalism should serve as a watchdog of powerful individuals and institutions; second, journalism should endeavor to discern truth from falsehoods; and third, journalism should provide a forum for substantive debate on matters of public interest and concern (McChesney, 2004, p. 57). Based on these journalistic standards, I contend domestic news coverage of the targeted killing program fails to meet the basic functions of a democratic press. Furthermore, I argue that deeply flawed and frequently deceptive news accounts of the targeted killing program promote an unrealistic and illusory picture of drone warfare—one that serves to legitimate and routinize this new kind of war.

In what follows, I identify three distinct but related shortcomings in domestic news coverage of America's drone wars: (1) uncritical reliance on official sources; (2) specious assessments of the accuracy and precision of drone strikes; and (3) a dearth of critical perspectives on the legal, ethical, and strategic implications of drone warfare. Throughout, I suggest these shortcomings undermine the American people's capacity to fully comprehend the consequences of the targeted killing program, thereby facilitating public acceptance of this immoral, shortsighted, and self-defeating foreign policy.

Taking my cue from Norman Solomon's *War Made Easy* (2005)—a historical analysis of the perception management techniques employed by government officials, and their enablers in the news industry, to promote US militarism and empire building—I aim to reveal the decisive role news discourse plays in normalizing and institutionalizing drone warfare. First, I examine the ideological function of language in reporting the US drone campaign. Here, I employ the linguistic concepts of *lexical categorization*, *transitivity*, and *nominalization* to demonstrate the significance of word choice and sentence structure in reflecting, reinforcing, and justifying unequal relations of power in America's drone wars.

Following this, I consider the relationship between journalistic *sourcing* practices and the remarkably sanitized depiction of drone warfare propagated through US corporate media: a portrayal that renders drone strikes an appealing military option for a majority of Americans. I hasten to add that my emphasis on sourcing practices is but one aspect of the sanitized depiction of robotic war. Despite an excess of graphic images documenting the damage drone strikes inflict on people and communities, this material rarely appears in US news media (Gregory, 2011; Scharrer & Blackburn, 2015; Stahl, 2013). The contradiction between the *surplus* of combat footage produced by this new kind of war and the *absence* of these images in news and official discourse may be one of the defining features of domestic coverage of the targeted killing program. All the more reason to attend closely the use of language in official statements regarding the "accuracy" and "precision" of US drone strikes.

Next, I explore the implications of the discursive strategy of *personalization* in domestic news accounts of drone strikes. On the one hand, the narrative emphasis on "eliminating" so-called high value targets is essential for constructing simple and replicable storylines that foster a (false) sense of achievement in America's decades-long war on terror. On the other hand, the triumphalism that accompanies the extrajudicial killing of suspected terrorists diverts attention away from the wholesale slaughter of innocent civilians. The chapter concludes with a brief discussion of an independent investigation of the targeted killing program that failed to gain traction in corporate news outlets; in large part, I want to suggest, because it challenges media narratives touting the efficacy of the US drone program.

Language and Power in Reporting the Drone Wars

As Pulitzer Prize-winning journalist Glenn Greenwald suggests in the epigraph that opens this chapter, "militant" is a keyword in domestic news coverage of the drone program. Synonyms such as "extremist," "jihadist," and "terrorist" abound in press accounts of drone strikes across the Muslim world; but the frequency with which the word militant is used in reporting America's drone wars constitutes what linguists call a lexical category. Categorization of this sort is, of course, a fundamental aspect of human language: "We manage the world, make sense of it, by categorizing phenomena, including people" (Fowler, 1991, p. 92). In other words, language and discourse are essential to the social construction of difference. Nevertheless, when lexical categories are employed in news discourse—a key site for the formation of public knowledge and opinion of social, political, and economic relations—they frequently work to construct and maintain unequal relations of power and authority. As Roger Fowler puts it, "categorization is a discursive basis for practices of *discrimination*" (p. 93).

None of which is to suggest that US news organizations are unique in this regard. For instance, in his analysis of the British Broadcasting Corporation's (BBC) coverage of the Palestinian-Israeli conflict, Leon Barkho (2010) found "the Palestinians that Israeli forces kill or injure, as well as their activities, fall invariably under the lexical category of 'militancy'" (p. 132). Barkho contends that the routine use of the word militant, in tandem with other pejoratives, is an instance of "over lexicalization"—or the repetitive and redundant use of different words to describe a specific concept or entity. In news discourse, over lexicalization serves to reinforce and legitimate dominant meanings; in the BBC's case, meanings that equate Palestinians with violent extremism. Lexical categories such as militant play a crucial role in supporting and justifying discrimination based on religious, ethnic, and political differences. It is here, first and foremost, in the discursive strategy of labeling the targets of US drone strikes as militants, that we can begin to appreciate the ideological effect of domestic news accounts of the drone wars.

In the context of America's war on terror, the term militant serves to construct whole groups of people as enemy Others: a diffused and determined adversary violently opposed to "our" way of life and ruthlessly engaged in a clash of civilizations that requires swift and decisive action on the part of the United States and its (predominantly) Western allies. Used in conjunction with a host of ideologically charged epithets—Islamic jihadist, Muslim extremist and the like—the phrase renders targets of US drone strikes as nameless and faceless abstractions. Without putting too fine a point on it, domestic news accounts of the targeted killing program employ the word with such regularity that it serves as a catchall term for *anyone* killed in a drone strike. Periodically, journalists qualify their reports with phrases like "suspected of being" and "thought to be" militants. All of which suggests that news workers—like the civilian leaders, intelligence analysts, and remote operators who authorize and conduct these lethal assaults—consistently fail to differentiate between armed insurgents, alleged terrorists, and innocent civilians.

Consider the following example taken from a 344-word story published in America's newspaper of record, the New York *Times*:

> An American drone strike in northwestern Pakistan killed at least four people suspected of being **militants**, Pakistani officials said Sunday … On Wednesday, a drone strike in Datta Khel, in the adjacent North Waziristan tribal region, killed at least 10 people who were thought to be **militants** … The Pakistani military is engaged in an offensive against **militants** in North Waziristan … The Pakistan military claimed success in its campaign against **militants** in South Waziristan after a military offensive in 2009, but the **militants** still have a presence in the region … American officials emphasize that remotely piloted drones are an essential weapon in the hunt for **militants** in areas where ground troops cannot operate. (Masood, 2014)

Through sheer repetition and frequent equivocation news workers routinize and condone indiscriminate killing under the guise of technological superiority and political expediency. Accordingly, this lexical category functions to promote and legitimate drone warfare for the American public. As we shall see, this ideological work is most apparent in news coverage of one of the most lethal and controversial aspects of the US drone program: so-called signature strikes.

Like word choice, sentence structure serves to legitimate and naturalize drone warfare in news discourse. The concepts of transitivity and nominalization—essential linguistic tools for representation and the construction of meaning from experience—are especially useful for establishing relationships between agents and action in written and spoken expression. As an analytical device, transitivity reveals how language functions to represent, evaluate, and classify specific actions. Depending upon the transitive formation involved, clauses represent the same action or event in very different ways. In doing so, transitives construct and reflect different interpretations of an action or event: interpretations that frequently maintain and reproduce unequal relations of power. For example, the phrase "soldiers attack militants" differs significantly from the phrase "militants attacked" insofar as the first phrase attributes the action "attack" to a set of agents "soldiers." The second phrase fails to identify the perpetrator of the action.

Fowler, Hodge, Kress, and Trew (1979) note that in proper English usage, placing the subject at the beginning of a clause, such as a headline, is vital for clarifying *who does what to whom*. Consider the following: "CIA Kills Al Qaeda's No. 2 Leader" (*Wall Street Journal*, June 17, 2015). In this rare instance—made all the more unusual by the unambiguous attribution for the lethal action to the American spy agency—the headline conforms to the three Ws rule: one which makes clear who did what to whom. But as discourse analysts have demonstrated repeatedly, reporters and headline editors frequently eschew this structure. As a result, the agent responsible for a particular action receives less attention or thematic prominence in news accounts. Put differently, failure to follow the three Ws rule of correct English usage obscures the relationship between agency and action, or deletes agency altogether. For example, the headline "6 Killed in Drone Strike" (*New York Times*, February 15, 2010) foregrounds the (anonymous) victims of an assault while leaving unsaid who was behind the attack. In other words, the headline fails to identify the agent responsible for this action. The use of transitives in headlines consistently serves to obscure or delete agency in news accounts of US drone strikes.

Another syntactic transformation common in the English language, nominalization, refers to a process whereby verbs are transformed into nouns or noun phrases. In addition to deleting agency, this transformation turns processes or events into agents or entities. Doing so, these nouns or noun phrases serve to reify abstract concepts as agents (Billig, 2008). Here, the new lexical term "drone strike" obscures the complex socio-technical system involved in surveilling, targeting, and firing upon suspected terrorists. Indeed, by some estimates, "a typical 'combat patrol' may have as many as 186 individuals working on it" (Chatterjee, 2015). Thus, in the headline "Drone Strike Kills Six Suspected Al-Qaeda Militants in Yemen" (*Newsweek*, February 3, 2016)—a geographically diffuse assault conducted by ground crews, remote pilots, sensor operators, and intelligence analysts using a sophisticated global network of data recording, transmission, and processing technologies—is transformed into a subject capable of taking action. Used in tandem with passive sentences, nominalization omits or conceals the agents responsible for a specific action.

Consider the following: "US Drone Strike in Pakistan Is Said to Have Killed 6 Militants" (*New York Times*, January 5, 2015). The complex noun phrase "US drone strike" transforms an action or process into an agent. Moreover, the passive construction "is said to have killed 6 militants" obscures and mitigates responsibility for the missile attack. In short, word choice, transitivity and nominal patterns represent events and situations from a particular point of view or perspective. Used in the study of news discourse, these concepts reveal, "how the details of texts can serve to reproduce the workings of ideology" (Billig, 2008, p. 784). Upon closer examination of US newspaper headlines, we can discern the ideological function of transitivity, nominalization, and passivization in domestic news coverage of the drone wars.

> **3 killed in Pakistan following US missile strikes**
> (*AP*, October 3, 2010)
>
> **Drone strike kills 21 suspected militants in northern Pakistan**
> (*Pittsburgh Tribune Review*, August 11, 2011)
>
> **US Drone Strikes Are Reported to Kill at Least 18 in Pakistan**
> (*New York Times*, June 7, 2011)

The *AP* report does not follow the three Ws rule associated with proper English usage. Instead, the headline foregrounds the death of three unidentified people and employs the complex noun phrase "US missile strikes" to attribute the action to American armed forces. Nonetheless, by inserting this nominal at the end of the clause, the headline significantly reduces the thematic prominence of the agent responsible for the lethal assault. In contrast, the Pittsburgh *Tribune Review* headline follows the three Ws rule, but does so by using the nominal "drone strike," thereby transforming a process into an agent responsible for the death of 21 "suspected" militants. Unlike the previous example, however, this headline does not use a modifier to attribute the assault to US forces. Finally, the New York *Times* headline attributes an assault on an unidentified group of "at least" 18 people using the complex noun phrase "US Drone Strikes," but employs a passive construction, "are reported," to obscure, rather than clarify, who does what to whom. Thus, word choice and sentence structure not only conceal the perpetrators behind these actions, but also trivialize the causalities of these assaults by failing to clearly identify those responsible.

In addition to concealing perpetrators and trivializing casualties through the use of transitivity and passivization, headline writers employ nominals to reify the complex socio-technical processes used for conducting drone strikes. As a result, headlines routinely assign agency directly to these killing machines. Consider the following examples:

> **Drone activity in Pakistan moves northward, official says**
>
> (*CNN*, January 6, 2010)
>
> **Drone kills fiend in Hebdo case**
>
> (*New York Daily News*, May 8, 2015)
>
> **Drones hunt Islamic State leaders; US launches secret campaign in Syria with help of CIA**
>
> (*Dayton Daily News*, September 2, 2015)

Notwithstanding dire predictions surrounding the development of autonomous weapon systems discussed in Chapter 1, characterizing drone strikes in this fashion further obscures responsibility, let alone accountability, for these lethal assaults. Instead, drones appear to operate on their on accord. In the CNN headline, "drone activity," suggests that these killing machines are a force of nature, or perhaps that they have taken it upon themselves to expand the war on terror to the northern regions of Pakistan. Doing so, this headline obfuscates the fundamental and decisive role policymakers play in expanding the drone campaign into undeclared battlefields. Conversely, drones take on a vengeful character in the New York *Daily News* account of an attack that killed a suspect in the slaughter of journalists from *Charlie Hebdo*, the French satirical magazine, in May 2015. Finally, the CIA's role in covert military operations in Syria is far less prominent in the Dayton *Daily News* headline than the weapon systems deployed by the American intelligence service. Rather than highlighting the transformation of the US intelligence agency into a paramilitary organization, the headline anthropomorphizes drones. Here, then, these killing machines single-mindedly pursue their quarry in yet another undeclared battlefield. Like the previous examples, this final headline portrays drones not only as active, but sentient beings capable of autonomous action.

All of which is to suggest that US news discourse routinely employs word choice and sentence structure to obscure rather than illuminate the conduct and character of America's drone wars. Of course, news workers are not alone in mystifying or concealing the inner-workings, let alone the consequences, of the US drone campaign. Official discourse invariably employs word choice and sentence structure of the sort described above (Jackson, 2005). These statements are then uncritically repeated in news discourse. This should come as no surprise, as journalists frequently take their cues from official sources—political leaders, Pentagon brass, military analysts and the like—whose statements are rarely questioned or challenged by journalists who frequently share basic ideological assumptions with their sources (Hallin, 1989). As a result, news discourse draws upon, echoes, and amplifies official discourse in important and decisive fashion. An examination of sourcing practices explains a great deal about the highly sanitized and upbeat assessment of drone warfare presented in US corporate media.

Officials Say ...

As noted in Chapter 2, reporters' uncritical reliance on official sources plays a crucial role in shaping news coverage of the American drone campaign. While such dependence is typical of journalistic practice in other domains, from economic and political journalism to crime and sports reporting, during wartime journalistic deference to official sources has serious implications for the veracity of public information regarding the causes, conduct, and potential consequences of the war effort (Solomon, 2005). The perceived quality and accuracy of this information invariably shapes public opinion. As a result, public understanding of a discrete military operation, such as a drone strike that "takes out" a suspected terrorist leader, rests in large measure on official assessments relayed through news channels. More generally, public perception of the legality, morality, and efficiency of the targeted killing program relies on the mediated discourse of just war, surgical strikes, and precision-guided munitions articulated by political and military officials.

To be clear, war reporting is fraught with hazards and difficulties. In addition to the usual impediments to fair and accurate reporting from warzones—the enormous financial resources and logistical support required for such reporting, perilous working conditions, and restricted access to battlefields—reporters face significant obstacles covering drone strikes in some of the world's most remote and increasingly inhospitable regions (McKelvey, 2011). They are not, however, insurmountable. For instance, the non-profit, Bureau of Investigative Journalism compiles and publishes complete data sets on US drone strikes in Pakistan, Yemen, and Somalia. And as we shall see in Chapter 7, non-governmental organizations such as Human Rights Watch and Amnesty International routinely document civilian casualties of the targeted killing program—albeit with scant mention in mainstream news media (Project Censored, 2015). As Code Pink co-founder Medea Benjamin observes, "corporate media outlets are more interested in running unverified government spin as 'news' than talking to people on the ground to actually uncover the truth" (Benjamin, 2013, p. 106).

Equally important, the secrecy surrounding the US drone campaign, coupled with the Obama administration's unprecedented crackdown on investigative journalists and whistleblowers, had a chilling effect on substantive, critical reporting of the drone wars (Harris, 2012). In short, this new kind of war exacerbates news workers' usual tendency to rely on official sources. We can detect this journalistic predilection in the following headlines from the Associated Press, the leading American wire service.

> **Officials: 2 drone strikes kill 13 in NW Pakistan**
> (*AP* September 29, 2009)
> **Officials: Suspected US strike kills 7 in Pakistan**
> (*AP* January 6, 2010)

At first glance, the AP headlines appear to provide readers with clear and concise information regarding military operations in Pakistan. Both attribute this information to "officials" who ostensibly have detailed information regarding the lethal assaults. Thus, the information presented in these headlines appears authoritative. Nonetheless, the headlines are rather ambiguous insofar as the identity of the officials making these statements is unclear. The body of both stories clarifies the matter—somewhat. The first report cites, "two Pakistani intelligence officials [speaking] on condition of anonymity because they were not authorized to release the information" (Afzal, 2009). Further down in the story, additional details are attributed to these same unnamed sources: "Seven insurgents died, they said, also on condition of anonymity, citing policy" (Afzal). The second report fails to explicitly identify the nationality of these unnamed officials altogether; the local authorities are Pakistani security personnel only by inference. Even a cursory examination of news accounts of US drone strikes reveals that this approach is the rule, not the exception. Simply put then, news accounts of the drone wars routinely repeat official claims based on (undisclosed) conditions of anonymity—conditions that are a matter of (classified) government and military policy. This lack of transparency further confounds the already difficult task of producing accurate and credible accounts of casualties in America's drone wars.

None of which is to suggest that this practice is unique to news coverage of the US drone campaign. As a matter of course, journalists and official sources routinely make informal arrangements that ensure reporters' access to knowledgeable speakers—provided that those sources are not publicly identified. "Journalism, as a profession, has a long, sordid history of misusing and abusing anonymous sources" (Richardson, 2016). As a consequence, news organizations attempt to mitigate the ethical shortcomings and transparency issues engendered by this technique through institutional guidelines and recommended "best practices." While this arrangement benefits reporters and their anonymous sources alike, this discursive practice has considerable advantages for government and military officials. As Maija Stenvall observes: "Even anonymous sources are usually construed as 'elite,' as possessing an authoritative, official status" (2010, p. 79). And while this practice typically yields ambiguous and highly circumscribed details regarding the particulars of any given drone strike, anonymity effectively shields officials from public scrutiny and accountability. All the while advancing a sanitized and reassuring, but wholly unrealistic and illusory picture of drone warfare.

Returning to the two examples above, both headlines provide precise casualty figures, but the dead are anonymous. Here, then, these headlines reflect but fail to highlight a despicable yet unacknowledged dimension of drone warfare: neither the officials nor the journalists who rely upon these anonymous sources really know who has been killed in a US drone strike. And while the first headline employs the nominal "drone strike" without clearly identifying the agent behind the attack, the second employs an even more complex and mystificatory noun phrase "Suspected US strike"—a clause that scarcely clarifies who did what to whom.

Of course, headlines reveal only a limited amount of information by design; readers must consult the body of the story for further details and, perhaps, analysis of the events depicted in these headlines. But upon closer examination, news headlines frequently obscure rather than reveal relevant information. For instance, in the following story from the New York *Times*, the headline editor reverses the order employed in the previous examples; thereby foregrounding an assertion and later attributing this claim to unnamed officials. Subsequent information relayed in the lead paragraph suggests a possible explanation behind this construction.

> **Strike Aimed at Militant in Pakistan, Officials Say**
> At least 10 people were killed in an American drone strike in northwest Pakistan on Thursday morning that was aimed at the leader of the Pakistani Taliban, according to Pakistani security officials. (Oppel & Tavernise, 2010)

The headline employs the lexical category militant to characterize the target of a US drone strike. However, the headline fails to mention a crucial detail: "at least" 10 people perished in the lethal assault. In stark contrast to headlines that routinely offer a precise number of casualties killed in such attacks—insofar as the numbers supplied by anonymous sources can be said to accurately reflect the carnage inflicted by missiles fired from US drones[1]—the *Times* inexplicably omits this vital information. Here, then, the attribution, "Officials Say" to the claim that a militant was the target of the deadly assault deflects attention away from the victims: people who may or may not have been associated with the intended target. That these victims are not described as militants, as is usually the case, further suggests that they were perhaps innocent bystanders in a botched targeted assassination. Finally, while these unnamed officials claim the air strike *aimed* at a known terrorist, the story reveals that the intended target "was not among the casualties."

Here, as elsewhere, journalists' use of anonymous sources frequently works "in a prescriptive rather than a normative fashion and are reactive in nature, narrowly tailored to correct the latest embarrassing mistake" (Richardson). Viewed in this light, the *Times*' headline and subsequent reliance on unnamed sources serves to conceal the loss of innocent lives while simultaneously maintaining the fiction of a precise, accurate, and effective targeted killing program. The balance of the story focuses on the intended target: Hakimullah Mehsud, a Pakistani Taliban leader, who orchestrated the 2009 assault on a CIA base in Afghanistan that killed seven Americans. The victims of the errant assault are neither identified nor mentioned again.

Personality Strikes and the Anonymous Dead

In November 2013, nearly four years after the fatal assault described above missed its intended target, Hakimullah Mehsud, the Taliban leader was killed in a US drone strike. The insurgent commander died in what war planners euphemistically describe as a "personality strike." News accounts of such attacks, which target specific individuals the US government identifies as terrorists, adhere to much the same narrative style and structure as the reports on the death of Anwar al-Awlaki discussed in Chapter 3. That is to say, news stories detailing the death of high profile terrorists follow a simple and easily replicable format: one that rhetorically constructs the targets of personality strikes as one-dimensional and therefore easily disposable (and replaceable) official enemies whose highly circumscribed histories neglect the social, political, and ideological factors behind their insurgencies.

Typically, these reports begin with a dramatic lead, based on statements made by unnamed official sources, describing the target and the location of the assault, followed by a succinct review the criminal allegations leveled against the suspected terrorist. These stories also recount the dead man's leadership position within a particular organization, occasionally speculating on his possible successor. Invariably, these news stories include after action assessments from official sources, or their surrogates in the news industry, who offer insights into what this latest drone strike might mean in the fight against the Taliban, or al-Shabab, or al-Qaeda, or ISIS or whatever insurgency is the terror group du jour. Regardless of the particulars, the calmly reassuring and distinctly celebratory tone of these reports is hard to miss. In this way, news accounts of personality strikes not only reflect the interdependent relationship between journalists and official sources, but also reveal the ideological work of news discourse surrounding the extrajudicial killing of so-called high value targets—HVTs in military parlance.

An analysis of a CBS Evening News account of Hakimullah Mehsud's death exposes the political and ideological significance of personalizing high value targets through television news discourse. CBS News anchor and managing editor, Scott Pelley, begins the story in dramatic fashion: "This may be the biggest day at the CIA since the killing of Osama bin Laden" (CBS News, 2013). As Pelley relates news of Mehsud's death, based on "multiple reports tonight," an over-the-shoulder graphic zooms in on a still image of the insurgent leader. From the very outset, then, this news story discursively equates the relatively unknown Mehsud with bin Laden, arguably the world's most infamous terrorist. Doing so, the report heightens the significance of this latest targeted assassination while contextualizing the assault in relation to America's ongoing war on terror. Pelley's lead-in to CBS's State Department Correspondent Margaret Brennan's report identifies Mehsud as the leader of the Pakistani Taliban and highlights his role in the death of "seven CIA employees." Subtly, but unmistakably, Pelley's opening remark and subsequent description suggests a sense of retribution over the deadly assault on a CIA station in Afghanistan years earlier. One pertinent detail missing in this account:

the prominent role that same CIA station played in the US drone campaign on the Afghan-Pakistan border (Gopal, Gorman, & Dreazen, 2010).

Brennan's voice over begins: "Sources say that Mehsud was killed in northwestern Pakistan." Thus, this "field report," filed not from some far-off battlefield, but from the safety and security of the Department of State in Washington DC, relies exclusively on official statements provided by unnamed sources. Another motion graphic, this one a full-screen map of the region, followed by a rapid zoom in to the location of the drone strike in Pakistan's semi-autonomous tribal area, Waziristan, operates in ideological fashion to underscore America's technological superiority and its unrivaled ability to project force into the most remote corners of the globe. File footage of the dead man employs a spot light to single out Mehsud from a crowd of followers—a visual technique that further contributes to the personalization cum demonization of the Taliban leader. Brennan's voice over continues: "Believed to be in his mid-thirties, Mehsud was considered one of the most wanted terrorists in the world. The U.S. government had a five-million-dollar bounty on his head" (CBS News).

This uncredited file footage yields to another full-screen motion graphic. Reinforcing the notion of America's global dominance, a world map overlays a computer readout of indeterminate letters and digits, suggesting the intelligence-gathering prowess of US counterterrorism forces. In turn, a "wanted poster" published by the State Department's Rewards for Justice program[2] wipes across the world map. The document includes yet another image of Mehsud along with biographical information and physical attributes including place of birth, nationality, known aliases, as well as skin tone, hair and eye color. Curiously, Brennan's initial statement indicating that Mehsud is "believed to be in his mid-thirties" belies the completeness of the dossier presented in graphic form. Without stating it as such, this admission undermines oft-repeated claims that US counterterrorism forces know precisely who it is they are targeting in lethal drone strikes.

Brennan's narration continues with a litany of Mehsud's crimes, including his claim to have orchestrated the failed 2010 Times Square bombing, as well as his indictment for the aforementioned assault on a CIA station. File footage from New York City, Afghanistan, and a funeral service, ostensibly for one of the seven CIA employees killed in the 2009 attack, reinforces the significance of the drone strike that ended Mehsud's life. The report then cuts to a recent BBC interview with Mehsud. While his words are not translated, Brennan notes, "Mehsud said, he'd continue to target Americans." Here, domestic reporters' proclivity for sound bite journalism meets their reluctance to translate statements made by official enemies—a technique that denies viewers much needed insight into the motives behind terrorist attacks and further obscures the historical grievances of insurgent movements. Brennan's synopsis of Mehsud's statement, with its vague yet ominous warnings of future terror plots targeting Americans, further contributes to a report whose tone and tenor unabashedly confirms the value of retributive justice for the admittedly heinous crimes committed by the Taliban leader.

The report ends with a "stand up" of Brennan, who appears on camera for the first time, in front of a row of flags, presumably from a press room in the US State Department. Brennan's closing remarks neatly encapsulate the role American broadcast news plays in legitimating and normalizing drone warfare—even in the face of widespread condemnation of the United States' flagrant violation of Pakistani national sovereignty and international law. Brennan concludes, "Just last week the Pakistani prime minister publicly asked President Obama to end all US drone strikes. But Scott, they clearly remain a very useful counter-terror tool" (CBS News). After years of acquiescence to the US drone campaign along the porous Afghan-Pakistan border, the Pakistani government openly denounced the controversial drone campaign. Nonetheless, Brennan, clearly supporting and affirming the position of American officialdom, offers anchor Scott Pelley, and millions of television viewers, confident assurances of the efficacy of US drone warfare.

As Fowler notes: "The obsession with persons, and the media's use of them as symbols, avoids serious discussion and explanation of underlying social and economic factors" (1991, p. 16). American reporters' fascination with high-value targets, a preoccupation shared with and cultivated by government officials, clearly operates in just this fashion. Willfully ignorant of the root causes of Islamic terrorism, the American press corps dutifully and triumphantly celebrates the death of the latest insurgent leader to perish in a US drone strike, and then quickly moves on to the next high value target. A tendency neatly captured by a CBS News headline, "Which al Qaeda terrorist should US target next?" (CBS, June 16, 2015). The ideological effects of this preoccupation with high profile terrorists are plain to see:

> The intense focus on [high value targets] who by any measure had a damnable lack of respect for human life, prevents any questioning of the tactics and their impact on less dastardly people—like innocent men, women, and children. (Benjamin, p. 109)

In short, the discursive strategy of personalization serves to justify and promote drone warfare by providing a fearful and acquiescent American public with a misleading sense of accomplishment in a decades-long war on terror.

But what of the anonymous victims of US drone strikes? As a report from the Inter Press Service (IPS) noted, residents in Pakistan's Federally Administered Tribal Areas (FATA) challenge official accounts of the number and identity of "militants" killed in American drone strikes, "people living in the affected areas are now questioning these figures, asking why they never get to know the names of the militants or see the bodies" (Yusufzai, 2013). In contrast to personality strikes, so-called signature strikes do not target known individuals. Instead, these assaults target individuals and groups whose actions and behaviors, based on "pattern of life" analysis obtained through signals intelligence from satellites, surveillance drones, and cell phone intercepts, suggest they may be armed insurgents (Hudson, 2015).

Despite repeated assurances that signature strikes rarely kill innocent civilians—in June 2011, Obama's counterterrorism advisor, John Brennan, famously claimed that for nearly a year, "there hasn't been a single collateral death because of the exceptional proficiency, precision of the capabilities we've been able to develop"[3]—this tactic continues to draw criticism from human rights organizations, international relations experts, as well as survivors and victims' family members, who condemn the practice as indiscriminate murder; albeit with scant attention in US corporate media. In point of fact, domestic news accounts rarely contradict official assessments of this tactic (Hudson). More critically, as Jeffery Bachman (2017) has demonstrated, America's preeminent newspapers, the New York *Times* and Washington *Post*, not only fail to accurately report civilian causalities, but also minimize the broader impact of the drone campaign on local communities.

Consider the restrained, casual, and occasionally flippant manner in which American journalists discussed civilian deaths in a December 2013 drone strike in Yemen. For instance, in a New York *Times* story headlined "Drone Strike in Yemen Hits Wedding Convoy, Killing 11" the unidentified casualties "appeared to be people suspected of being militants linked to al Qaeda, according to tribal leaders in the area, but there were also reports that several civilians had been killed" (Worth, 2013). Like the stories discussed above, this account equivocates as to the identity of the dead. What's more, the story provides background information—read justification—for the errant attack by recalling the previous week's AQAP assault on the Yemen Defense Ministry. Citing reports on a local website, *Bayda Press*, which published the names of 11 dead men "said to be closely associated with al-Qaeda in the area," the *Times* story refuses to follow suit. Instead, the balance of the story focuses on "the dilemma" President Obama faced in light recent assurances he made regarding the "near-certainty that noncombatants will not be injured or killed"[4] in US drone strikes.

The following day, NBC Nightly News broadcast a brief update on the Yemen strike:

> The death toll now appears to be 15 with five more injured from a US drone strike in Yemen that missed its target this week and struck what witnesses described as a wedding party. It was the twenty-fourth reported drone strike in Yemen so for this year. (NBC, 2013)

Notwithstanding the rather unusual inclusion of "witness" testimony in corporate news accounts of US drone strikes, as NBC News anchor Brian Williams reads this short item an over the shoulder graphic features file footage of drones taxiing on a runway and in flight, below a caption reads "DEADLY ATTACK." Visually, this brief story celebrates American technological superiority while studiously avoiding images that might convey the carnage unleashed in this, the latest deadly drone assault on Yemeni civilians. Doing so, NBC upholds a time honored tradition: American television news revels in showing off American fire power, but rarely shows what happens to "people on the receiving end" of these sublime weapons (Solomon, 2005, p. 121).

Earlier that same day, the tabloid New York *Post* ran one of the more objectionable, yet revealing headlines in the annals of the drone wars: "BRIDE AND BOOM! US drone strike in Yemen hits 'Qaeda' wedding party" (*New York Post*, December 13, 2013). While such callous reporting is rare, the unapologetic jingoism implicit in this headline remains a fixture in domestic news coverage of the targeted killing program. Put differently, the consequences of "collateral damage"—Pentagon-speak for innocent civilians killed in US military assaults—remains unseen and unexamined, and purposefully so, in what might accurately be described as an exercise in perception management designed to simultaneously minimize and justify the grim reality of the American drone campaign. In subsequent chapters, we explore the brutality of drone warfare from two unique perspectives: that of victims and survivors of US drone strikes (Chapter 7); and that of military personnel who have served in the drone wars and have since emerged as some of the fiercest critics of America's targeted killing program (Chapter 9).

***

On October 15, 2015 *The Intercept*—an online news site co-founded by investigative journalists Glenn Greenwald, Laura Poitras, and Jeremy Scahill—published an exposé detailing the inner workings of the Obama administration's targeted killing program (Scahill et al., 2015). Based on classified documents obtained by an unnamed US intelligence source, "The Drone Papers" offers fresh insights into America's secretive drone campaign in Afghanistan, Yemen, and Somalia. Despite, or more accurately perhaps, because of the disturbing details contained in "The Drone Papers," corporate news outlets all but ignored *The Intercept's* investigative series.

Billed as the most in-depth look at the US drone assassination program to date, the exposé documents fundamental flaws in the targeted killing program, from an over reliance on faulty signals intelligence and overstatements of the precision of US drone strikes to lowballing civilian causalities: so-called collateral damage estimates (CDE) in the Orwellian language of what lead author Jeremy Scahill calls "the bureaucracy of assassination." When asked about White House reaction to *The Intercept's* series, Scahill wryly noted that officials refused to comment on "purported internal documents" (Goodman & Gonzalez, 2015). Such evasions are to be expected inside the beltway. Nevertheless, it is troubling how readily corporate news outlets likewise refused to address the stunning revelations contained in the leaked documents described in "The Drone Papers." After all, the series and its provenance have all the makings of a blockbuster news story: classified documents, a government whistleblower, and details of abuses and atrocities surrounding one of the most secretive military operations in US history.

While "The Drone Papers" presents a searing indictment of the criminality of the US drone campaign, the media blackout surrounding this explosive report reveals the shortcomings of an all too compliant, and frequently complicit, press corps. Willfully ignorant of *The Intercept's* stunning revelations, American news workers continue to legitimate the United States' repeated and flagrant violations of international law. What passes for substantive reporting of the targeted killing program is neatly summed up in the proverb "see no evil, hear no evil, speak no evil."

Notes

1. Human rights groups and others, such as the Bureau of Investigative Journalism, regularly challenge official estimates of civilian casualties in US drone strikes. Significantly, when

these disputes arise, news workers frequently temper this criticism. Note the passive construction of this headline, "CIA Claim of No Civilian Deaths from Drones Is Disputed" (*New York Times*, 11 August 2011).

2. According to its website, the Rewards for Justice program, sponsored by the US Department of State, "continues to be one of the most valuable U.S. Government assets in the fight against international terrorism." Available: https://www.rewardsforjustice.net/english/index.html.
3. See Coll (2014).
4. See "Remarks by the President at the National Defense University," White House.gov, https://www.whitehouse.gov/the-press-office/2013/05/23/remarks-president-national-defense-university.

References

Afzal, H. (2009, September 29). Officials: 2 drone strikes kill 13 in NW Pakistan. *Associated Press. Lexis-Nexis Universe: General News Topics.*

Bachman, J. (2017). The *New York Times* and *Washington Post*: Misleading the public about US drone strikes. *Journalism Studies*, *18*(4), 470–494.

Barkho, L. (2010). The BBC's discursive strategy and practices vis-à-vis the Palestinian-Israeli Conflict. In J. E. Richardson (Ed.), *Language and journalism* (pp. 127–143). London: Routledge.

Benjamin, M. (2013). *Drone warfare: Killing by remote control.* London: Verso.

Billig, M. (2008). The language of critical discourse analysis: The case of nominalization. *Discourse & Society*, *19*(6), 783–800.

CBS News. (2013, November 1). Officials: US drone kills Pakistani Taliban leader. *CBS Evening News*. Retrieved from http://www.cbsnews.com/news/officials-us-drone-kills-pakistani-taliban-leader/

Chatterjee, P. (2015, July 12). No lone rangers in drone warfare. *TomDispatch.* Retrieved from http://www.tomdispatch.com/blog/176022/tomgram:_pratap_chatterjee,_no_lone_rangers_in_drone_warfare

Coll, S. (2014, November 24). The unblinking stare. *The New Yorker*. Retrieved from http://www.newyorker.com/magazine/2014/11/24/unblinking-stare

Fowler, R. (1991). *Language in the news: Discourse and ideology in the press*. London: Routledge.

Fowler, R., Hodge, B., Kress, G., & Trew, T. (1979). *Language and control.* London: Routledge.

Goodman, A., & Gonzalez, J. (2015, October 16). "The drone papers" reveals how faulty intel & secret "kill chain" mark suspects, civilians for death. *Democracy Now!* Retrieved from https://www.democracynow.org/2015/10/16/precisely_wrong_faulty_intelligence_drives_civilian

Gopal, A., Gorman, S., & Dreazen, Y. J. (2010, January 4). Taliban member claims retaliation. *Wall Street Journal.* Retrieved from http://www.wsj.com/articles/SB126246258911313617

Gregory, D. (2011). From a view to a kill: Drones and late modern war. *Theory, Culture & Society*, *28*(7–8), 188–215.

Hallin, D. C. (1989). *The "uncensored war": The media and Vietnam*. Berkeley, CA: University of California Press.
Harris, P. (2012, April 11). The Obama administration's disturbing treatment of whistleblowers. *The Guardian*. Retrieved from http://www.theguardian.com/commentisfree/cifamerica/2012/apr/11/obama-administration-whistleblowers-paul-harris
Hudson, A. (2015). Thanks to reliance on "signature" drone strikes, US military doesn't know who it's killing. *Truthout*. Retrieved from http://www.truth-out.org/news/item/32166-thanks-to-reliance-on-signature-drone-strikes-us-military-doesn-t-know-who-it-s-killing
Jackson, R. (2005). *Writing the war on terrorism: Language, politics and counter-terrorism*. Manchester: Manchester University Press.
Masood, S. (2014, September 28). In Pakistan's tribal north, US drone strike kills four. *New York Times*. Retrieved from https://www.nytimes.com/2014/09/29/world/in-pakistans-tribal-north-us-drone-strike-kills-four.html?_r=0
McChesney, R. W. (2004). *The problem of the media: US communication politics in the 21st century*. New York, NY: Monthly Review Press.
McKelvey, T. (2011). Covering Obama's secret war. *Columbia Journalism Review*. Retrieved from http://www.cjr.org/feature/covering_obamas_secret_war.php
NBC News. (2013, December 13). Yemen drone strike hits wedding party, witnesses say. *NBC Nightly News*. Retrieved from http://www.nbcnews.com/video/nightly-news/53824889#53824889
Oppel, R, A., Jr., & Tavernise, S. (2010, January 15). Strike aimed at militant in Pakistan, officials say. *New York Times*. *Lexis-Nexis Universe: General News Topics*.
Project Censored. (2015, December 7). Corporate news sources fail to fully report US drone strike casualties. *Project Censored*. Retrieved from http://www.projectcensored.org/corporate-news-sources-fail-to-fully-report-us-drone-strike-causalities/
Richardson, R. (2016). Journalism's dark matter. *FAIR*. Retrieved from http://fair.org/home/journalisms-dark-matter/
Scahill, J., Begley, J., Currier, C., Devereaux, R., Mass, P., Gallagher, R., & Turse, N. (2015). The drone papers. *The Intercept*. Retrieved from https://theintercept.com/drone-papers/
Scharrer, E., & Blackburn, G. (2015). Images of injury: Graphic news visuals' effects on attitudes toward the use of unmanned drones. *Mass Communication and Society*, *18*, 799–820.
Solomon, N. (2005). *War made easy: How presidents and pundits keep spinning us to death*. Hoboken, NJ: John Wiley & Sons.
Stahl, R. (2013). What the drone saw: The cultural optics of the unmanned war. *Australian Journal of International Affairs*, *67*(5), 658–674.
Stenvall, M. (2010). Unnamed sources as rhetorical constructs in news agency reports. In J. E. Richardson, J. E. (Ed.), *Language and journalism* (pp. 78–92). London: Routledge.
Worth, R. F. (2013, December 12). Drone strike in Yemen hits wedding convoy, killing 11. *New York Times*. Retrieved from http://www.nytimes.com/2013/12/13/world/middleeast/drone-strike-in-yemen-hits-wedding-convoy-killing-11.html
Yusufzai, A. (2013). Drone killings show numbers, not bodies. *Inter Press Service*. Retrieved from http://www.ipsnews.net/2013/12/drone-killings-show-numbers-bodies/

PART III

WITNESSING

· 7 ·

SURVIVORS SPEAK

I wasn't scared of drones before, but now when they fly overhead I wonder, will I be next?
—Nabila Rehman, 9-year-old survivor of a US drone strike

On July 1, 2016, the Obama administration released its highly anticipated report on civilian casualties of the US drone program. According to figures provided by the Office of the Director of National Intelligence, from 64 to 116 "non-combatants" perished in airstrikes between January 2009 and December 31, 2015.[1] Over that same period, the report added, counterterrorism operations killed as many as 2500 "combatants." The US government estimate represents a fraction of the number of civilian deaths documented by the Bureau of Investigative Journalism, Amnesty International, and Human Rights Watch (Bachman, 2017). What's more, the statement failed to mention by name Pakistan, Yemen, Somalia, and Libya—countries where the attacks occurred. Nor did the three-page report include specific information, such as the victims' names, their country of origin, and the date and location of the lethal assaults; let alone how the government defines "combatants" and selects individuals for targeted assassination. According to NPR, the Obama administration stopped short of offering an apology for civilian casualties but claims "it takes responsibility and, where possible, it makes payments to victims' families" (Horsley, 2016). At the same time, President Obama issued an executive order calling for greater safeguards to protect civilians and requiring his successor to provide a yearly accounting of civilian deaths in the *undeclared* battlefields of America's war on terror. It remains to be seen if President Donald Trump will follow suit or, as he indicated throughout the 2016 campaign, he will repeal this and other executive orders issued by Barack Obama.

The White House announcement and the attendant press coverage reveal a good deal about the discursive strategies and practices of politicians and news workers alike. Consider the timing of the administration's report. Government officials routinely release embarrassing or controversial information on weekends and holidays, when news production and consumption decline: all the better to manage public perception. Not coincidentally, then, the Obama administration released its estimate of civilian causalities on a Friday afternoon, prior to the Independence Day holiday. Only a handful of news reports acknowledged the significance of the report's timing. The New York *Times* noted rather cryptically, "In a seeming acknowledgment that the long-awaited disclosure would be greeted with skepticism by critics of the drone program, the administration issued the numbers on a Friday afternoon before a holiday weekend" (Savage & Shane, 2016). By contrast, the independent news site, *The Intercept*, put the report's release in broader context, noting that the White House made its announcement on the eve of the Fourth of July celebration, "after seven years of selective leaks and official secrecy" (Devereaux, 2016).

Corporate media dutifully echoed administration talking points—invariably attributed to unnamed sources. For instance, an *AP* story published in advance of the White House announcement noted that officials "spoke on condition of anonymity because they weren't authorized to disclose the information" (Riechmann & Hennessey, 2016). The *Times* cited "a senior administration official, speaking on condition of anonymity" who claimed the causality figures and executive order were designed to "bolster public support [for the drone program] at home and abroad" (Savage & Shane). Finally, by way of explaining dramatic discrepancies between government estimates and independent reports on civilian casualties, the Washington *Post* quoted "one of several senior officials who briefed reporters under White House anonymity rules" (DeYoung & Miller, 2016). The contradiction between Obama's claims of transparency and his administration's sourcing guidelines went unremarked by the White House press corps.

Notwithstanding the *Times*' suggestion that releasing the report before a holiday weekend would somehow blunt critics' reaction to the administration's figures, the majority of press reports did, in fact, include responses from human rights organizations and other independent observers. For instance, the aforementioned *AP* wire story included reaction from Federico Borello of the Washington, DC-based Center for Civilians in Conflict. The *Times*' own story also included comments from Borello, as well as Jameel Jaffer of the ACLU, and Bill Roggio, editor of *The Long War Journal*. As the *Post* put it: "Release of the non-combatant casualties counts provoked a flood of reaction from groups who questioned the value of the numbers" (DeYoung & Miller). Significantly, while US corporate media offered drone critics an opportunity to challenge the government's estimate, these stories invariably toed the administration's line; frequently citing the White House's claim that its figures, based on classified intelligence, are superior to those produced by rights groups, whose numbers may be "tainted by terrorist propaganda" (Savage & Shane).

Independent news outlets were far more skeptical of White House spin. For instance, a story published by *Common Dreams* describes a report issued by Reprieve, a UK-based human rights organization (Prupis, 2016). Released in advance of the government report, the study titled "Opaque Transparency"[2] highlights previous statements that undermine and contradict the administration's casualty estimates; such as John Brennan's June 2011 assertion that, "For the past year, there hasn't been a single collateral death" in Pakistan; and President Obama's May 2013 assurance that drones strikes are conducted only when there is "near certainty of no civilian casualties." Conversely, a story written by Naureen Shah, Director of Security and Human Rights at Amnesty International, welcomes the administration's efforts, calling it a "remarkable shift" in the conduct of the secretive targeted killing program, but expresses doubt over the government's figures: "the drone data could be entirely misleading—and provide a veneer of legitimacy to unlawful killings" (Shah, 2016). Originally published in *The Guardian* and reprinted by *Common Dreams*, this analysis concludes on a cautionary note.

> The communities that live on the other end of drone fire already suffer from a peculiar invisibility in the US public's understanding. … Raw numbers, like today's, could contribute to that erasure: we have numbers to assign to the people who were killed, and now the government has assigned categories to them, but no names. Without more, theirs are just the lives of others. (Shah)

This chapter—the first of three that employ the concept of "media witnessing"—features the names, voices, and perspectives of drone survivors. Their stories constitute an *alternative order of discourse* to the dominant and domesticating orders of discourse examined in previous chapters. Such testimony constitutes an alternative discourse precisely because victim and survivor stories rarely appear in corporate news media—lest the American people begin to question the efficacy, let alone the morality, of this new kind of war.

> Think about it. Have you ever seen an image of a bleeding survivor or a charred drone victim in our media, much less the victims' devastated homes and villages? That is no accident. (Pierson, 2013)

In contrast, independent journalism—news, opinion, and analysis produced and distributed "outside of traditional corporate systems and news organizations"[3]—provide a resource and a venue for media witnessing in the drone wars. Here, then, I want to highlight the significance of witnessing as a "discursive practice" (Peters, 2011): one that reveals the brutality of drone warfare otherwise concealed from public view. Doing so, I contend the deliberate and purposeful exclusion of survivors' stories make corporate news organizations complicit in legitimating indiscriminate violence and normalizing US war crimes. In turn, corporate media silence surrounding drone victims and survivors contributes to the American people's ignorance—and tacit approval—of an illegal, immoral, and counterproductive counterterrorism strategy.

The chapter proceeds with an analysis of news coverage of the Rehman family of Pakistan's historic testimony before the US Congress on the drone strike that killed their beloved matriarch, Momina Bibi. Following this, I examine the decisive role legal researchers and independent news outlets (both foreign and domestic) play in breaking the silence and relaying the experience of drone survivors to American audiences. The chapter concludes with a discussion of firsthand accounts of the drone wars from an American journalist, a Pakistani activist, and Yemeni teenager. Taken together, these stories offer a stark contrast—what I'm calling an alternative order of discourse—to news and official discourse that habitually celebrates the accuracy and precision of US drone strikes.

Not a Militant, But My Mother

As noted in Chapter 3, the year 2013 was a watershed in the public life of drones. The leaked Justice Department memo outlining the Obama administration's legal justification for executing Anwar al-Awlaki and Senator Rand Paul's filibuster brought unprecedented attention and scrutiny to the targeted killing program. In October of that year, Rafiq ur Rehman, a schoolteacher from Tabbi, North Waziristan—a remote village in Pakistan's Federally Administered Tribal Area (FATA)—traveled to Washington DC with his children, Zubair, 13 and Nabila, 9. There the family gave first-of-its kind testimony on civilian causalities of the drone program at an informal hearing co-sponsored by Congressman Alan Grayson (D-FL) and Brave New Foundation, a non-profit, progressive media organization. The briefing came amidst renewed criticism of America's drone program: days earlier Nawaz Sharif, the newly elected Pakistani Prime Minister, met with President Obama and urged him to end the drone campaign in his country; and a recently completed inquiry conducted by the UN Special Rapporteur on extrajudicial, summary or arbitrary executions questioned the legality of America's targeted killing program.

In anticipation of his historic testimony, Rafiq Rehman published an open letter to President Obama in *The Guardian* titled "Please tell me, Mr. President, why a US drone assassinated my mother?" The letter, which served as the basis for his testimony and was reprinted by a handful of independent news outlets, read in part:

> Nobody has ever told me why my mother was targeted that day. The media reported that the attack was on a car, but there is no road alongside my mother's house. Several reported the attack was on a house. But the missiles hit a nearby field, not a house. All reported that five militants were killed. Only one person was killed—a 67-year-old grandmother of nine. … Not a militant, but my mother. (Rehman, 2013)

In plain language and with clear frustration, Rehman questions the legal justification for the drone strike that ended his mother's life, and challenges media narratives that uncritically echo official statements made by local and distant authorities. A far cry from the US government's classified and highly circumscribed after action assessments, Rehman's letter speaks directly to the loss of innocent lives—and to the woeful lack of accountability in the drone wars. Like his family's testimony on Capitol Hill, however, Rafiq's letter received scant attention in US corporate media; a sharp contrast to extensive coverage in the independent and alternative press. Nevertheless, their witnessing offers a glimpse into the shattered lives across Pakistan—and throughout the Greater Middle East—wrought by America's decades-long war on terror. What follows draws directly from the Rehman family's testimony, as well as excerpts taken from published accounts of what was aptly described as a "grim first" (Sledge, 2013).

On October 24, 2012, the Rehman family was preparing to celebrate the Muslim holiday of Eid-al Adha: a particularly festive occasion for children. Nine-year-old Nabila told Pacifica radio's *Democracy Now!*

> I was outside with my grandmother, and she was teaching me how to tell the difference between okra that was ripe and not ripe. We were going to prepare it for our Eid dinner the next day. And then I had heard a dum-dum noise. Everything became dark. And I had seen two fireballs come down from the sky. (Goodman & Gonzalez, 2013)

Like his sister, Zubair was working in the field alongside his grandmother. He told lawmakers, "As I helped my grandmother in the field, I could see and hear the drone hovering overhead, but I didn't worry. Why would I worry? Neither my grandmother nor I were militants." Just then a missile struck Momina Bibi, killing her instantly. Returning home from the market, Rafiq realized something was horribly wrong when he noticed a freshly dug grave nearby. Speaking to reporters, he explained,

> I had asked some little children who they were preparing the burial for. And they said, "This is Latif Rehman's mother." And that's my older brother. So I knew at that point that my mother had been killed by an American drone. (Goodman & Gonzalez, 2013)

Rafiq went on to describe how villagers kept him from seeing his mother's decimated remains. Meanwhile, his children were taken to distant hospitals. Nabila's arm was badly cut, but Zubair's wounds required multiple surgeries. The boy told lawmakers, "We spent the night in great agony in the hospital and the next morning I was operated on. That is how we spent Eid."

In an era of smartphones and viral videos these oral accounts may seem quaint, and, perhaps, less convincing than dramatic footage of some calamity or another. But given the virtual blackout of such footage in US corporate media—despite the surfeit of graphic imagery produced through the sophisticated optics onboard these killing machines (Gregory, 2011; Stahl, 2013)—such testimony offers all too rare but compelling insight into the brutality of the American drone campaign. As Frosh and Pinchevski observe, "'Being there' matters … the event is singular, and its witnesses are forever irreplaceable in their privileged relation to it" (2011, p. 7). Despite, or more accurately perhaps, because of their privileged relation to that day's events, Zubair and

Figure 7.1: Nabila Rehman looks on as her brother, Zubair, testifies before a Congressional committee on civilian casualties of the US drone campaign. *Photo credit: Jason Reed. Source: Reuters.*

Nabila's testimony failed to gain much traction– at least with corporate news organizations. The same media institutions that clamor for exclusive interviews with witnesses to unusual or spectacular events were curiously indifferent to these firsthand accounts of a lethal drone attack. Herein, we can detect the US news media's double standard of "worthy" and "unworthy" victims of atrocity (Herman & Chomsky, 1988). As Murtaza Hussain observed on *Al Jazeera English*, the US press corps' indifference to Nabila's plight offers a revealing contrast to American news workers' fascination with Malala Yousafzai—another Pakistani girl scarred by a violent assault.

> While Malala was feted by Western media figures, politicians and civic leaders for her heroism, Nabila has become simply another one of the millions of nameless, faceless people who have had their lives destroyed over the past decade of American wars. (Hussain, 2013)

Equally disturbing, only five out of 430 members of the US Congress—all Democrats—attended the special hearing (McCauley, 2013). Had they the courage to attend, lawmakers might have learned of the emotional and psychological trauma besetting the Rehman family—the sort of dramatic testimony congressional representatives instinctively gravitate toward, especially during televised hearings and reelection campaigns. Asked about the consequences of the attack, Nabila told *Democracy Now!* "Ever since the strike, I'm just scared. I'm always scared. All of us little kids, we're just scared to go outside" (Goodman & Gonzalez, 2013). Likewise, Zubair offered keen insights into the pernicious psychological effects of drone warfare: "I no longer prefer blue skies. In fact, I now prefer gray skies. The drones do not fly when the skies are gray. When the sky brightens, drones return and we live in fear." Zubair's eloquence made headlines in the alternative press, but failed to receive much attention in US corporate media (Figure 7.1).

When Rafiq spoke of his loss, he likened his mother to a string of pearls that held the family together. But it was his inability to comprehend the injustice of his mother's death that moved his translator to tears. Speaking of his passionate commitment to education, he asked lawmakers:

> But how do I teach something like this? How do I explain what I myself do not understand? How can I in good faith reassure the children that the drone will not come back and kill them, too, if I do not understand why it killed my mother and injured my children?

In the aftermath of tragedy, politicians and pundits habitually drone on about the need to "learn lessons." But in the absence of the political will to hear the stories of drone survivors, and in the midst of the media silence surrounding their broken lives, such lessons are few and far between. Speaking to *Al Jazeera America*, Rafiq Rehman said he wanted to make the American people aware of the horrors of drone warfare—and enlist their help in putting an end to it. "That's why I came here—I thought if they heard my story, they would want to listen to me and influence their politicians" (Khan, 2013). But between lawmakers' indifference and the news media silence, relatively few Americans had the opportunity to hear the Rehman family's story. Tellingly, as the Rehmans gave their congressional testimony, President Obama was meeting with the CEOs of Lockheed Martin and Northrop Grumman—two of the world's leading weapons manufacturers (Hanly, 2013).

Other Voices, Other Drones

Idris Farid, eyewitness to a March 17, 2011 drone strike that killed over 40 people in Datta Khel, North Waziristan, told investigators: "Everything was devastated. There were pieces—body pieces—lying around. There was lots of flesh and blood." The Obama administration claimed the attack targeted militants. But witnesses and survivors said the assault on a tribal council meeting, called a jirga, took the lives of innocent civilians. Firoz Ali Khan relates an equally graphic account of the drone strike that killed his father-in-law: "These missiles are very powerful. ... Whatever is left is just little pieces of bodies and cloth." Documented in a 2012 Stanford-NYU law report, "Living Under Drones,"[4] these stories challenge what the report refers to as "the dominant narrative" of this new kind of war: a storyline that depicts drone warfare as an accurate, precise, and effective counterterrorism measure. The study's blunt assessment: "This narrative is false." Based on nine months of research, including over 130 interviews with survivors, local aid workers, legal experts and others, the study employs eyewitness testimony and forensic investigation to refute longstanding US government claims that the targeted killing program causes minimal harm or injury to civilian populations.

While corporate news outlets made only passing reference to the study,[5] independent news organizations proved invaluable for disseminating the report's findings to wider publics. Doing so, audiences of the alternative press received a rigorous, eye-opening, and damning account of the civilian costs of drone warfare: a counter-narrative to the sanitized depiction of drone warfare propagated by official Washington and US corporate media. The divergent coverage of "Living Under Drones" in mainstream and independent news outlets underscores a fundamental aspect of media witnessing: "There is clearly an institutional politics of contemporary media witnessing that informs how witnessed worlds are represented and shared and who may depict them and appear in them" (Frosh & Pinchevski, 2011, p. 11). Based on the dearth of reporting on the Stanford-NYU study, it seems corporate news organizations find eyewitness accounts of US drone strikes neither compelling nor especially newsworthy. At best, the mainstream media's refusal to discuss the study's findings reflects institutional subservience to political and military power; at worst, it reveals willful ignorance of US war crimes. Conversely, by disseminating the report's findings and relaying survivors' stories, independent media bear witness to the wanton criminality of drone warfare, thereby opening up a discursive space to challenge this deeply flawed and morally reprehensible counterterrorism policy.

To that end, the report's lead investigators, human rights attorneys James Cavallaro and Sarah Knuckey, discussed their findings with independent news outlets. For instance, in an interview with *Truthout*, a digital platform for investigative journalism and political analysis, Knuckey explained that the report is "intended to insert the stories of those who bear the brunt of US drone strike policies—civilians in North Waziristan—into US public and policy making debates" (Knefel, 2012). Significantly, these stories not only attest to the carnage of drone warfare, but also reveal the debilitating fear produced by incessant drone surveillance. As one Waziri man put it, "The drones are having a really bad effect on this area. Everyone is under so much psychological pressure that many have to go to the city to see a doctor." The report concludes that America's drone program "terrorizes men, women, and children, giving rise to anxiety and psychological trauma among civilian communities" (International Human Rights, 2012).

Furthermore, "Living Under Drones" documents for the first time a practice known as a "double-tap," wherein a drone fires on a target, loiters in the area for several minutes, and then fires a second time as first responders arrive on the scene. "The result is that locals trying to provide help, and sometimes official humanitarian aid workers, are hit in the follow-up attacks" (Knefel). Western leaders and journalists routinely condemn terrorist organizations that use this same tactic to inflict maximum harm on so-called soft targets: civilians, first-responders and the like. However, when US military and paramilitary units use drones in similar fashion, politicians and news workers remain silent. Citing recent statements made by UN Special Rapporteur Christof Heyns, the study contends these practices constitute war crimes. And yet, despite such allegations, corporate media all but ignored the report. It was left to independent and progressive news organizations to break the silence surrounding the study, its findings, and its implications for US foreign and national security policy.

One of the more innovative approaches to "popularizing" this research (Vannini, 2012) involved close collaboration between legal scholars and the aforementioned Brave New Foundation. Through its affiliated media unit, Brave New Films (BNF), the foundation produced and distributed an online video to coincide with the September 2012 release of the Stanford-NYU study. Featuring interviews with witnesses, survivors, and the report's lead authors, the video condenses an exhaustive 165-page research document into a riveting 7-minute clip.[6] Recording and distributing eyewitness testimony for wider publics in this fashion, BNF's viral video typifies media witnessing, or what Frosh and Pinchevski define as "witnessing performed *in*, *by* and *through* the media. … the systematic and ongoing reporting of experiences and realities of distant others to *mass audiences*" (2011, p. 1). This is but one instance of

BNF's "media intervention" (Howley, 2013) into an otherwise anemic public debate over the US drone program; the organization's hour-long documentary film *Unmanned: America's Drone Wars* was screened prior to the Rehman family's testimony before Congress.

Like the Stanford-NYU study, subsequent independent investigations of civilian causalities in US drone strikes generated scant attention in mainstream news outlets. While investigations issued by the likes of Amnesty International and Human Rights Watch may generate headlines for a day or two, there is precious little follow up or analysis of these studies in corporate media. On the other hand, independent news organizations offer extensive coverage of these studies, and prove essential for raising public awareness of mounting civilian casualties in America's expanding drone program. Consider the lethal assault on a Yemeni wedding party discussed in Chapter 6. In December 2013, an operation conducted by the Joint Special Operations Command (JSOC) targeted a wedding procession making its way to a village on the outskirts of Yemen's central city, Rad'a. At the time, corporate news outlets offered ambiguous accounts of a drone strike that, as NBC stated, "missed its target" (NBC, 2013). Despite reports of civilian casualties—a violation of the Obama administration's new rules to protect non-combatants—the story failed to gain much traction beyond the 24/7 news cycle.

Two months later, Human Rights Watch issued a report "A Wedding that Became a Funeral" on the signature strike that claimed 15 lives and injured 12. Only a handful of domestic news organizations covered the study's release, and those that did invariably framed the story in terms of the policy implications of the airstrike. Details of the report's findings, including eyewitness testimony of the lethal assault, were few and far between. In contrast, the report's author, Letta Tayler, appeared on *Democracy Now!* to discuss her findings. The segment opens with videotape testimony of Abdullah Muhammad al-Tisi describing the drone strike that killed his son, Ali, and a dozen more men. The wedding procession of "12 or 15 cars with 60 or 70 men on board" was returning to the groom's village following a traditional meal with the bride's tribe when the assault began.

> A drone was hovering overhead all morning. There were one or two of them. One of the missiles hit the car. The car was totally burned. Four other cars were also struck. When we stopped, we heard the drone fire. Blood was everywhere, and the people killed and injured were scattered everywhere. The area was full of blood, dead bodies and injured people. I was injured. I saw the missile hit the vehicle behind the car my son was driving. (Goodman & Gonzalez, 2014)

Here again, an eyewitness to a US drone strike provides graphic details of the ensuing carnage—it is a vivid account that contradicts the dominant narrative articulated through news and official discourse. Doing so, survivors not only challenge the veracity of published accounts of drone strikes, but also, critically, invite audiences to reevaluate their understanding of this new kind of war, its victims, and its consequences. Thus, the discursive practice of witnessing involves "stating one's experience for the benefit of an audience that was not present at the event and yet must make some kind of judgment about it" (Peters, p. 25). Of course, this evaluative dimension of witnessing may explain the virtual blackout of survivors' stories in corporate news media. After all, as Norman Solomon (2005) argues, sanitized coverage of US military operations is standard operating procedure for American news workers—and essential for ensuring public consent to a foreign policy predicated on violence, intimidation, and the use of force.

In contrast to the official statements and selective leaks that constitute a false narrative of the "pinpoint accuracy" and "surgical precision" of US drone strikes, survivors speak with undeniable moral and cultural authority: a perspective largely absent from news and official discourse, let alone public and policy debates over the drone program. "Witnessing in this sense," Peters contends, "suggests a morally justified individual who speaks out against unjust power" (p. 30). Viewed in this light, Abdullah Muhammad al-Tisi's witnessing speaks truth to power. His stark conclusion is difficult to deny, much less ignore: "The US government made a big mistake. They killed innocent people. This was a serious crime. They turned many kids into orphans, many wives into widows" (Goodman & Gonzalez, 2014).

Firsthand Accounts

Independent investigators, like Letta Tayler of Human Rights Watch, bear witness to drone strikes using a variety of research methods and recording techniques. Describing her work on *Democracy Now!* Tayler explained:

> This was my seventh or eighth trip to Yemen in recent years, many of those trips to look at this particular issue of targeted killings. I met with relatives and family members there, as well as government officials, academics, journalists and so forth. The most compelling testimony, of course, was from the family members. (Goodman & Gonzalez, 2014)

Investigations of this sort occur without benefit of the researcher's firsthand experience, and well after the relevant events have transpired. Put differently, these investigations are removed in time, but present in space at the site of an atrocity. As such, human rights attorneys, like reporters more generally, serve as *mediators* for the eyewitness accounts of others—a topic taken up in greater detail in the next chapter. Here, I focus instead on firsthand accounts from individuals who have experienced the sublime terror of drone strikes for themselves, and who have subsequently shared their experience with distant audiences. Such accounts underscore the fundamental character of witnessing as a discursive practice: the transformation of experience into language (Peters, p. 27).

One of the first such accounts comes from New York *Times* correspondent David Rohde. In November 2008, the Pakistani Taliban kidnapped Rohde and two Afghan colleagues. They spent seven months as hostages in Pakistan's FATA region before making their escape. Rohde's account of his experience living under drones echoes those documented in the independent investigations discussed above. But with two crucial differences: first, Rohde relates his dramatic experience in his own words; second, and more important, the names and stories documented in those reports are largely absent from the pages of the *Times*, America's newspaper of record.

> It was March 25, and for months the drones had been a terrifying presence. Remotely piloted, propeller-driven airplanes, they could easily be heard as they circled overhead for hours. To the naked eye, they were small dots in the sky. But their missiles had a range of several miles. We know we could be immolated without warning. (Rohde, 2009)

Here, Rohde bears witness to the terror of drone warfare, not merely as journalist relating the experience of others, but rather as a survivor of a lethal assault. Apart from war correspondents, journalists rarely put themselves in mortal danger on a regular basis. But as Peters reminds us, "To bear witness is to put one's body on the line. Within every witness, perhaps, stands a martyr, the will to corroborate words with something beyond them, pain and death being the last resorts" (p. 30). That said, reporters occupy a privileged position in war zones; they can come and go as they please, unlike combatants and local populations with little hope of escaping the carnage—apart from death. Nonetheless, news workers who witness atrocity and experience trauma, as Rohde and his colleagues had, speak with much the same moral and cultural authority as any survivor (Tait, 2011). And in the intervening years, Rohde emerged as a vocal critic of the US drone program (2012, 2013). That reporters can also profit from their experience—as Rohde has with the publication of a book about his ordeal—reflects unequal relations of cultural capital in bearing witness to drone warfare.

Like so many victims and survivors of US drone strikes, David Rohde was not deliberately targeted for assassination. Few individuals selected for extrajudicial execution live to talk about it. Malik Jalal is an exception to the rule. Writing in *The Independent*, Jalal, a leader of the North Waziristan Peace Committee (NWPC), describes his unique circumstance straight away:

> I am in the strange position of knowing that I am on the "Kill List." I know this because I have been told, and I know because I have been targeted for death over and over again. Four times missiles have been fired at me. I am extraordinarily fortunate to be alive. (Jalal, 2016)

Jalal's harrowing tale of four near misses over a 15-month period—assaults that killed or gravely injured family and friends—makes for dramatic reading, and further undermines the US government's specious claims regarding the accuracy and precision of its targeted killing program. Equal parts pragmatism, resilience, and defiance, Jalal's response to his precarious situation underscores the insidious effects of drone warfare on the daily lives of individuals, families, and entire communities.

> I soon began to park any vehicle far from my destination, to avoid making it a target. My friends began to decline my invitations, afraid that dinner might be interrupted by a missile. I took to the habit of sleeping under the trees, well above my home, to avoid acting as a magnet of death for my whole family.

Jalal further explains why he traveled to Britain to tell his story: "I decided that if Westerners wanted to kill me without bothering to come speak with me first, perhaps I should come to speak to them." Thus, Malik Jalal's column, much like Rafiq Rehman's open letter to Barack Obama, highlights another crucial dimension of witnessing: the exchange of knowledge and experience. "Witnessing presupposes a discrepancy between the ignorance of one person and the knowledge of another: it is an intensification of the problem of communication more generally" (Peters, p. 26). Herein lies the social, cultural, and political significance of witnessing in the drone wars. As we've noted throughout this book, news and official discourse surrounding the US drone program constructs a false narrative—a sanitized depiction of a brutal tactic that takes the lives of innocent people, terrorizes local communities, and fuels animosity toward the United States and its allies. Survivors like Jalal offer a compelling counter-narrative: one that creates a discursive space to challenge this misleading narrative; raises public awareness of the atrocities being perpetrated in our names; and provides a forum to debate this increasingly common if legally suspect and morally reprehensible approach to counterterrorism. In the absence of this sort of communication and debate, the prospect of armed drones terrorizing whole populations the world over is all but certain. I explore the paradoxical relationship between global communication and the advent of robotic warfare in the book's concluding chapter.

While Malik Jalal's letter employs an element of gallows humor to convey the horrors of drone warfare, Mohammed Tuaiman's story highlights the power of pathos in communicating the human costs and consequences of living (and dying) under drones. In October 2011, Mohammed's father, Saleh, and 17-year-old brother, Jalil, were killed in a US drone strike in Yemen. The deaths of the family's eldest males put enormous pressure on the young boys to support their struggling family. Three years later, Mohammed spoke with journalists about his family's loss at the hands of Americans and their "death machines." Mohammed told reporters, "In their eyes, we don't deserve to live like people in the rest of the world and we don't have feelings or emotions or cry or feel pain like all the other humans around the world" (Madlena, Patchett, & Shamsan, 2015). The teenager's words underscore the value of media witnessing for promoting empathy and understanding of the pain and suffering of distant others (Katz, 2011).

On January 26, 2015, just months after speaking with Western journalists, a CIA drone strike killed 13-year-old Mohammed and two other "suspected" terrorists (Greenwald, 2015). In the months before he was killed, Mohammed used the video camera reporters gave him to document his daily life in al Zur village in Yemen's Marib province. Featuring scenes of children at play, young boys tending their camels, and family members huddled around a TV set, the footage includes interviews Mohammed conducted with his siblings. This raw material provides a vivid firsthand account of the consequences of the US drone program on the lives and fortunes of the Tuaiman clan. Upon learning of his death, *The Guardian* collected Mohammed's footage and produced a haunting portrait of a family coping with loss, financial hardship, and incessant fear of American drones.[7]

Standing on the shore of a lake where Saleh Tuaiman frequently brought his young boys to swim, Mohammed's brother fondly recalls such outings. "I remember we came to this place together. ... He used to teach us how to pray and how to grow crops such as corn." In another scene, Mohammed asks his younger brother, "Who killed your dad?" Sitting astride a camel, the boy coolly responds: "America." But perhaps the most revealing and poignant moment comes when Mohammed asks the same questions of his little sister, Shaima. The young girl's delight with speaking to her brother on camera quickly turns to despair, then furious indignation.

MOHAMMED: Where is your dad?
SHAIMA: Heaven.
MOHAMMED: Why?
SHAIMA: He was killed by an American drone.
MOHAMMED: Why did it kill him?
SHAIMA: Because of treachery.
MOHAMMED: Why did it kill him?
SHAIMA: He was killed by treachery.

This remarkable exchange—an example of the experiential interview, a broadcast news staple—is virtually nonexistent in domestic news coverage of the drone wars. That it took a 13-year-old boy from tribal Yemen, using a camera borrowed from British journalists, to bring this story to American audiences exemplifies the double standards of US news organizations. That is to say, victims of violence perpetrated by "official enemies" like al Qaeda and ISIS receive sustained and sympathetic news coverage in American news outlets (Greenwald). In contrast, innocent victims of the US drone program rarely merit a mention, let alone empathetic treatment. As noted in Chapter 6, the media silence surrounding victims and survivors of the drone wars is purposeful; yet another perception management technique used to promote, justify, and perpetuate American militarism. But as we have seen throughout this chapter, the voices of survivors offer an alternative perspective on drone warfare: one that not only contradicts sanitized coverage of US drones strikes, but also, critically, engenders empathy for innocent victims of America's war on terror.

Faheem Qureshi was 13 years old when he and his family fell victim to a US drone strike: an assault Barack Obama ordered on the third day of his presidency. The attack took the lives of two uncles and a cousin, dashed Faheem's dreams of becoming a chemical engineer, and left the young man as the family's sole breadwinner. The intended target—a Taliban leader—escaped unharmed. Seven years later, *The Guardian* newspaper told Qureshi's story in an expose on innocent victims of the US drone program. As reported by Spencer Ackerman, the newspaper's national security editor, Faheem's life changed forever that day. Ackerman (2016) writes:

> It took nearly 40 days for Qureshi to emerge from a series of hospitals, all of which he spent in total darkness. Shrapnel had punctured his stomach. Lacerations covered much of his upper body. Doctors operated on the entire left side of his body, which had sustained burns, and used laser surgery to repair his right eye. They could not save his left.

Like the Rehman family and the other survivors we've met, Faheem is looking for answers. In this, his first interview with an American journalist, he told Ackerman that while he was in the hospital, he kept asking the same questions. "What did I do for which I was punished so badly? What did my family do? Why did it happen to me?" For victims of what political leaders and news workers routinely describe as "senseless violence" such questions are commonplace. In the era of killing by remote control, these questions take on global urgency.

Even so, American news organizations rarely see fit to tell these stories. Rather, as Glenn Greenwald points out, domestic news accounts serve "primarily as a tool the Obama administration has used to display its so-called strength in going against accused terrorists" (Knefel). In contrast, over the course of the decades-long war on terror, legal researchers and independent journalists have sought out survivors of US drone strikes to learn their stories and relate their experiences to an ill-informed and seemingly apathetic American public. Put another way, corporate news outlets typically depict drone warfare as a neat and efficient enterprise: a sublime exercise in American ingenuity and technological superiority. To this dominant discourse, survivors offer a stark alternative: one that reveals the sublime terror of the targeted killing program. In the following chapters, we continue our examination of this alternative order of discourse as it has been articulated by an eloquent youth activist from Yemen (Chapter 8) and American drone operators who defiantly oppose to this new kind of war (Chapter 9).

Notes

1. The White House Fact Sheet does not differentiate between attacks launched by manned and unmanned aircraft, and does not include airstrikes conducted in the "active war zones" of Afghanistan, Iraq and Syria. Available https://www.whitehouse.gov/the-press-office/2016/07/01/fact-sheet-executive-order-us-policy-pre-post-strike-measures-address.

2. See "The Truth About US Drone Strikes," *Reprieve*, http://www.reprieve.org.uk/wp-content/uploads/2016/06/Obama-Drones-transparency-FINAL.pdf.
3. See Park Center for Independent Media at Ithaca College for this distinction between independent and corporate media. Available at http://www.ithaca.edu/rhp/independent media/.
4. See "Living under drones: Death, injury, and trauma to civilians from US drone practices in Pakistan," https://law.stanford.edu/wp-content/uploads/sites/default/files/organization/149662/doc/slspublic/Stanford-NYU-LIVING-UNDER-DRONES.pdf.
5. For instance, CNN buried a reference to the report in a story about American activists joining Pakistani protests against the US drone program. See "American activists in Pakistan to protest US drone strikes," http://www.cnn.com/2012/10/05/world/asia/pakistan-us-drone-protest/.
6. See *Living Under Drones* [video], https://www.youtube.com/watch?v=6yMOzvmgVhc.
7. See "We dream about drones, said 13-year-old Yemini before his death in a CIA strike," *The Guardian*, https://www.theguardian.com/world/2015/feb/10/drones-dream-yemeni-teenager-mohammed-tuaiman-death-cia-strike?CMP=share_btn_tw.

References

Ackerman, K. (2016, January 23). Victim of Obama's first drone strike: "I am the living example of what drones are." *The Guardian*. Retrieved from https://www.theguardian.com/world/2016/jan/23/drone-strike-victim-barack-obama

Bachman, J. (2017). The *New York Times* and *Washington Post*: Misleading the public about US drone strikes. *Journalism Studies*, *18*(4), 470–494.

Devereaux, R. (2016, July 1). Obama administration finally releases its dubious drone death toll. *The Intercept*. Retrieved from https://theintercept.com/2016/07/01/obama-administration-finally-releases-its-dubious-drone-death-toll/

DeYoung, K., & Miller, G. (2016, July 1). White House releases its count of civilian deaths in counterterrorism operations under Obama. *Washington Post*. Retrieved from https://www.washingtonpost.com/world/national-security/white-house-releases-its-count-of-civilian-deaths-in-counterterrorism-operations-under-obama/2016/07/01/3196aa1e-3fa2-11e6-80bc-d06711fd2125_story.html

Frosh, P., & Pinchevski, A. (Eds.). (2011). *Media witnessing: Testimony in the age of mass communication*. New York, NY: Palgrave Macmillan.

Goodman, A., & Gonzalez, J. (2013, October 31). "Too scared to go outside": Family of Pakistani grandmother killed in US drone strike speaks out. *Democracy Now!* Retrieved from http://www.democracynow.org/2013/10/31/too_scared_to_go_outside_family

Goodman, A., & Gonzalez, J. (2014, February 21). Turning a wedding into a funeral: US drone strike in Yemen kills as many as 12 civilians. *Democracy Now!* Retrieved from http://www.democracynow.org/2014/2/21/turning_a_wedding_into_a_funeral

Greenwald, G. (2015, February 10). The US media and the 13-year-old Yemeni boy burned to death last month by a US drone. *The Intercept*. Retrieved from https://theintercept.com/2015/02/10/u-s-media-13-year-old-yemeni-boy-killed-u-s-drone/

Gregory, D. (2011). From a view to a kill: Drones and late modern war. *Theory, Culture & Society*, *28*(7–8), 188–215.

Hanly, S. (2013, November 1). Obama meets with drone makers during drone victims' testimony. *Digital Journal*. Retrieved from http://www.digitaljournal.com/article/361355

Herman, E., & Chomsky, N. (1988). *Manufacturing consent: The political economy of mass media*. New York, NY: Random House.

Horsley, S. (2016, July 1). White House documents number of civilians killed in US drone strikes. *NPR*. Retrieved from http://www.npr.org/2016/07/01/484381660/white-house-documents-number-of-civilians-killed-in-u-s-drone-strikes

Howley, K. (Ed.). (2013). *Media interventions*. New York, NY: Peter Lang.

Hussain, M. (2013, November 1). Malala and Nabila: Worlds apart. *Al Jazeera English*. Retrieved from http://www.aljazeera.com/indepth/opinion/2013/11/malala-nabila-worlds-apart-201311193857549913.html

International Human Rights and Conflict Resolution Clinic (Stanford Law School) & Global Justice Clinic (NYU School Of Law). (2012, September). *Living under drones: Death, injury, and trauma to civilians from US drone practices in Pakistan*. Retrieved from https://law.stanford.edu/wp-content/uploads/sites/default/files/organization/149662/doc/slspublic/Stanford-NYU-LIVING-UNDER-DRONES.pdf

Jalal, M. (2016, April 16). "I'm on the kill list": What if feels like to be hunted by drones. *The Independent*. Retrieved from http://www.independent.co.uk/voices/i-am-on-the-us-kill-list-this-is-what-it-feels-like-to-be-hunted-by-drones-a6980141.html

Katz, E. (2011). Preface to the paperback edition: No more excuses. In P. Frosh & A. Pinchevski (Eds.), *Media witnessing: Testimony in the age of mass communication* (pp. vii–viii). New York, NY: Palgrave Macmillan.

Khan, N. (2013, October 29). Drone victims give firsthand account of attack. *Al Jazeera America*. Retrieved from http://america.aljazeera.com/articles/2013/10/29/congressional-briefingonpakistanidronevictimsdraws5lawmakers.html

Knefel, J. (2012, September 25). "Whatever is left is just pieces of body and cloth": New report details the horror of living under drones. *Truthout*. Retrieved from http://www.truth-out.org/news/item/11756-whatever-is-left-is-just-pieces-of-bodies-and-cloth-new-report-details-the-horror-of-living-under-drones

Madlena, C., Patchett, H., & Shamsan, A. (2015, February 10). We dream about drones, said 13-year old Yemeni before his death in a CIA strike. *The Guardian*. Retrieved from https://www.theguardian.com/world/2015/feb/10/drones-dream-yemeni-teenager-mohammed-tuaiman-death-cia-strike

McCauley, L. (2013, October 29). Congressional no-show at "heart-breaking" drone survivor hearing. *Common Dreams*. Retrieved from http://www.commondreams.org/news/2013/10/29/congressional-no-show-heart-breaking-drone-survivor-hearing

NBC News. (2013, December 13). Yemen drone strike hits wedding party, witnesses say. *NBC Nightly News*. Retrieved from http://www.nbcnews.com/video/nightly-news/53824889#53824889

Peters, J. D. (2011). Witnessing. In P. Frosh & A. Pinchevski (Eds.), *Media witnessing: Testimony in the age of mass communication* (pp. 23–41). New York, NY: Palgrave Macmillan.

Pierson, C. (2013, October 1). US to drone victims: Shut up. *CounterPunch*. Retrieved from https://www.counterpunch.org/2013/10/01/us-to-drone-victims-shut-up/

Prupis, N. (2016, June 30). New report highlights US legacy of lies on civilian drone deaths. *Common Dreams*. Retrieved from http://www.commondreams.org/news/2016/06/30/new-report-highlights-us-legacy-lies-civilian-drone-deaths

Rehman, R. (2013, October 25). Please tell me, Mr. President, why a US drone assassinated my mother. *The Guardian*. Retrieved from https://www.theguardian.com/commentisfree/2013/oct/25/president-us-assassinated-mother

Riechmann, D., & Hennessey, K. (2016, July 1). AP sources: Obama to reveal civilian deaths from drones. *Associated Press*. Retrieved from http://hosted2.ap.org/apdefault/3d-281c11a96b4ad082fe88aa0db04305/Article_2016-07-01-US--Drones-Civilian%20Casualties/id-caa55acbd66d4b9ebc2017815f9e592a

Rohde, D. (2009, October 21). A drone strike and dwindling hope. *New York Times*. Retrieved from http://www.nytimes.com/2009/10/21/world/asia/21hostage.html?pagewanted=all&_r=0

Rohde, D. (2012, March 1). How Obama's drone war is backfiring. *Reuters*. Retrieved from http://blogs.reuters.com/david-rohde/2012/03/01/how-obamas-drone-war-is-backfiring/

Rohde, D. (2013, May 24). Obama's overdue step on drones. *Reuters*. Retrieved from http://blogs.reuters.com/david-rohde/2013/05/24/obamas-overdue-step-on-drones/

Savage, C., & Shane, S. (2016, July 1). US reveals death toll from airstrikes outside war zones. *New York Times*. Retrieved from http://www.nytimes.com/2016/07/02/world/us-reveals-death-toll-from-airstrikes-outside-of-war-zones.html?_r=0

Shah, N. (2016, July 1). When will the White House tell us the whole truth about drone killings? *Common Dreams*. Retrieved from http://www.commondreams.org/views/2016/07/01/when-will-white-house-tell-us-whole-truth-about-drone-killings?utm_campaign=shareaholic&utm_medium=printfriendly&utm_source=tool

Sledge, M. (2013, October 29). Why did America kill my mother? Pakistani drone victim comes to Congress for answer. *Huffington Post*. Retrieved from http://www.huffingtonpost.com/2013/10/29/pakistani-drone-victim-congress_n_4171975.html

Solomon, N. (2005). *War made easy: How presidents and pundits keep spinning us to death*. Hoboken, NJ: John Wiley & Sons.

Stahl, R. (2013). What the drone saw: The cultural optics of the unmanned war. *Australian Journal of International Affairs*, *67*(5), 658–674.

Tait, S. (2011). Bearing witness, journalism and moral responsibility. *Media, Culture & Society*, *33*(8), 1220–1235.

Vannini, P. (Ed.). (2012). *Popularizing research: Engaging new genres, media, and audiences*. New York, NY: Peter Lang.

· 8 ·

MR. AL-MUSLIMI GOES TO WASHINGTON

The drone strikes are the face of America to many Yemenis.

—Farea al-Muslimi

A months-long lull in the American drone campaign over Yemen came to an abrupt end in the skies above the remote mountain village of Wessab, some 9 hours drive from the Yemini capital Sana'a. The drone strike killed five people from the impoverished farming community. Days later, on April 23, 2013, Farea al-Muslimi, a youth activist and freelance journalist born and raised in Wessab, appeared before the US Senate's first-ever public hearing on drone warfare. Co-chaired by Senator Dick Durbin (D-IL) and ranking member Senator Ted Cruz (R-TX), the Senate Judiciary Committee hearing, "Drone Wars: The Constitutional and Counterterrorism Implications of Targeted Killing," featured testimony from legal experts, policy analysts, and retired military officials.

Despite mounting pressure for greater transparency, however, the Obama administration refused to send a representative to discuss the drone program with lawmakers. Chairman Durbin, a longtime ally of the President, framed the historic hearings this way: "The use of drones has, in stark terms, made targeted killing more efficient and less costly—in terms of American blood and treasure. There are, however, long-term consequences, especially when these airstrikes kill innocent civilians" (Durbin, 2013). In the last of the day's prepared statements, Farea al-Muslimi spoke directly to the consequences of drone warfare for the Yemeni people, and to the strategic implications of the targeted killing program for both Yemen and US national security.

Like the previous chapter, this discussion examines al-Muslimi's testimony through the lens of recent scholarship on media witnessing. Specifically, I draw on Ashuri and Pinchevski's (2011) analytical framework of "witnessing as a field" to better understand and appreciate the politics of bearing witness to drone warfare. Accordingly, the field of witnessing consists of three interdependent spheres or zones: eyewitnesses, mediators, and audiences. Throughout this discussion, I contend al-Muslimi's testimony and the attendant news coverage, such as it was, reveal the complex and multifaceted character of power/knowledge relations in America's drone wars. I begin with an examination of the discursive strategies Farea al-Muslimi employs in his statement before the Judiciary Committee. Doing so, I demonstrate how al-Muslimi's interdiscursive practice establishes trust—the "basic currency" (Ashuri & Pinchevski, p. 133) of media witnessing—between himself, his Senate interrogators, and the viewing audience. Next, I analyze the relatively sparse domestic news coverage of the hearings. Indeed, the virtual blackout of al-Muslimi's testimony in corporate media further reveals the news industry's complicity in normalizing and legitimating America's targeted killing program. Finally, and notwithstanding the paucity of news coverage of al-Muslimi's testimony, I situate this episode in light of the "moral obligations" (Katz, 2011, p. viii) that come with witnessing the suffering of distant others in, by, and through the media. In this way, I hope to underscore the decisive role media witnessing plays in the formation of an alternative order discourse on drone warfare.

A War of Mistakes

In contrast to the military, legal, and policy experts who preceded him at the Senate hearing, Farea al-Muslimi's youth, Muslim faith, and status as a foreign national complicate his ability to establish trust with his audience. At one point during his testimony, television coverage cut away for a reaction shot of a skeptical looking Senator Lindsey Graham (R-SC). Nonetheless, through a skillful blend of autobiography, investigative reporting, and policy analysis, al-Muslimi establishes his credibility, not only as a witness to scenes of atrocity, but also as a thoughtful critic of US counterterrorism strategy. Put differently, al-Muslimi's statement combines different discourses and genres to articulate an alternative perspective on the targeted killing program: one that challenges the dominant narrative celebrating the precision, accuracy, and efficiency of drone warfare. In this way, al-Muslimi's testimony illustrates the value and importance of interdiscursive practice for establishing trust with his (proximate) Senate interlocutors and the (distant) viewing audience of C-SPAN—the non-profit, public affairs cable network that carried the Judiciary Committee's historic hearing.[1]

After thanking co-chairs Durbin and Cruz for inviting him to speak, al-Muslimi tells lawmakers, "Just six days ago, my village was struck by an American drone in an attack that terrified the region's poor farmers" (C-SPAN, 2013). He goes on to describe his family's subsistence lifestyle and the lack of economic and educational opportunities for his people. Recounting his good fortune as a recipient of a US State Department scholarship, al-Muslimi relates his experience as an exchange student attending Rosamond High School in California, where he "learned about American culture, managed the school basketball team, and participated in trick or treat on Halloween" (C-SPAN). Al-Muslimi then describes his host in the United States: a US Air Force member. "Most of my year was spent with him and his family. He came to mosque with me and I went to church with him. And he became my best friend in America." Reflecting on his time in the United States, and in an effort to establish ethos with his American audience, al-Muslimi concludes, "I went to the US as an ambassador for Yemen, and I came back to Yemen as an ambassador for the US" (C-SPAN).

Through the genre of autobiography al-Muslimi establishes his credibility with committee members while simultaneously bridging the cultural divide between himself and the television audience. Viewed as a *discursive performance*, al-Muslimi's testimony leverages the cultural capital he acquired through his State Department scholarships, academic achievements, and professional reputation to win over both his proximate and distant audiences.

Despite his shock and horror at the American drone strike on his village, al-Muslimi's equanimity and eloquence enhance his trustworthiness. Without stating it as such, then, al-Muslimi's testimony invokes the discourse of "soft power"—an approach to international relations predicated on the use of economic and cultural influence, rather than coercion, intimidation and military force to achieve national security and foreign policy objectives (Nye, 1990). As evidenced by his stirring testimony, al-Muslimi's life story epitomizes the value of soft power for bridging cultural differences and establishing

Figure 8.1: Farea Al-Muslimi addresses the first-ever US Senate hearing on the targeted killing program. *Source*: C-SPAN.

constructive, mutually beneficial, and respectful relations between nations and peoples.

Al-Muslimi's discursive strategy combines this personal history with journalistic authority to further enhance his credibility. Describing his experience as a freelance journalist covering America's war on terror, al-Muslimi relates his intimate knowledge of the human costs and consequences of the targeted killing program.

> I have spoken to many victims of US drone strikes. Like a mother in Lahj, who had to identify her innocent 18-year-old son's body through a video on a stranger's cell phone. Or the father in Shaqra, who held his 4 and 6-year-old children as they died in his arms. (C-SPAN)

Unlike the overwhelming majority of his American colleagues, al-Muslimi's reporting does not rely on official statements and selective leaks. Rather, his investigations draw on personal visits to the sites of US drone strikes across Yemen, and features the voices of victims of the targeted killing program. Doing so, al-Muslimi bears witness to these atrocities and relays survivors' stories with a clarity and conviction sorely lacking in domestic news coverage of the drone wars (Figure 8.1). Equally important, al-Muslimi relates his first-hand experience with these killing machines, when, during a reporting trip to Abyan, he confronted the terrifying reality of living under drones.

> Suddenly, I heard the buzz. … My heart sank. I felt helpless. … I couldn't help but think that the operator of this drone just might be my American friend with whom I had the warmest and deepest friendship in America. … I was torn between the great country that I know and love and the drone above my head that could not differentiate between me and some AQAP militant. (C-SPAN)

As Ashuri and Pinchevski observe, reporters are integral components of an institutional apparatus that produces media witnesses. Put differently, journalists are agents who mediate—record, reproduce, and disseminate—the pain and suffering of "distant others to mass audiences" (Frosh & Pinchevski, 2011, p. 1) By the same token, however, in the course of their investigations, journalists are also "a species of eyewitness" (Ashuri & Pinchevski, p. 138). In the passage above, then, al-Muslimi assumes a dual role as both eyewitness to the insidious effects of US counterterrorism operations, and as a mediator for victims and survivors of the US drone campaign in Yemen. In short, al-Muslimi's credibility and status rests in large measure on his "being (from) there." Moreover, al-Muslimi's reportage underscores the potential risk and trauma associated with "bearing witness" to atrocity (Rentschler, 2011). Significantly, the risk al-Muslimi confronts in the course of his reporting comes not only from American military forces, but also from growing Yemeni resentment and hostility toward the United States. In his written statement to the Judiciary Committee, al-Muslimi claims that rising anger over the US drone program, "makes it dangerous for me to even acknowledge having visited America … It's sometimes dangerous to even admit I have American friends" (Al-Muslimi, 2013). Assuming such risk, al-Muslimi further demonstrates his trustworthiness for the assembled lawmakers and distant television audiences.

A third element of al-Muslimi's interdiscursive strategy draws upon a prominent trope in news and official discourse; namely, the use of drones to kill terror suspects when capture is deemed "infeasible"—an Obama-era euphemism for avoiding American casualties in counterterrorism operations. In al-Muslimi's estimation, the capture or kill quandary is grossly overstated. Worse yet, the fictitious sublimity of US drone strikes repeatedly hands AQAP propaganda victories—regardless of whether or not civilians are injured or killed in these assaults. Unlike his fellow panelists, then, whose technocratic jargon sometimes obscures more than it reveals, al-Muslimi articulates the tactical and strategic failures of the targeted killing program in plain language.

> My understanding is that a man named Hameed al-Radmi was the target of the drone strike. Many people in Wessab know al-Radmi and the Yemeni government could

> *easily* have found and arrested him. Al-Radmi was well known to government officials and even local government could have captured him if the US had told them to do so. (C-SPAN)

Here, al-Muslimi contradicts official statements on US counterterrorism operations in Yemen. As noted in Chapter 2, these accounts invariably highlight the close cooperation between American and Yemeni forces, and extol the value of precision strikes for eliminating violent extremists and avoiding civilian casualties. In contrast, al-Muslimi suggests that American forces act unilaterally, with little cooperation from either the Yemeni central government or local authorities. The result, al-Muslimi concludes, undermines American soft power and turns onetime allies into lifelong enemies.

> In the past, what Wessab's villagers knew of the US was based on my stories about my wonderful experiences here. The friendships and values I experienced and described to the villagers helped them to understand the America that I know and that I love. Now, however, when they think of America, they think of the terror they feel from the drones that hover over their heads, ready to fire missiles at any time. (C-SPAN)

A far cry from winning hearts and minds, US counterterrorism strategy in Yemen, including but not limited to drone strikes,[2] terrorize local populations, weaken popular support for the central government, and drive victims' families to join AQAP's ranks. In sum, al-Muslimi's testimony offers an alternative perspective on drone warfare: one wholly at odds with the dominant narrative rendered in news and official discourse.

All of which underscores the political character of the field of witnessing. As Ashuri and Pinchevski argue, "witnessing is to be regarded as subject to contest and struggle" (p. 135). We can detect this struggle between competing agents in the field of witnessing during the question and answer portion of the Senate hearings. For instance, addressing the issue of blowback raised by Senator Al Franken (D-MINN), retired Air Force Colonel Martha McSally offers standard issue assurances of the technological sophistication and bureaucratic checks of the drone program. At one point, McSally tells lawmakers: "A remotely piloted aircraft actually gives you better precision with a small warhead, with persistence overhead, with the ability to abort at the last minute, with the whole chain of command and the lawyers watching" (C-SPAN). In turn, al-Muslimi exposes the strategic fallacy behind America's targeted killing program.

SENATOR AL FRANKEN:	What I wonder is ... this new type of warfare ... Is it a different kind of blowback? Is there a different kind of reaction, because of the very nature of it?
FAREA AL-MUSLIMI:	Yes. I think the main difference between with this is that it adds into Al Qaeda propaganda—that Yemen is in a war with the United States. The problem of Al Qaeda, if you look at the war effort in Yemen, it's a war of mistakes. The less mistakes you make, the more you win. And the drones have simply made more mistakes than AQAP has ever done—in the matter of civilians. And you're also neglecting a very simple fact, which is you actually can capture this person. It's not impossible. (C-SPAN)

Challenging the dominant narrative surrounding the targeted killing program, al-Muslimi's testimony articulates an alternative perspective that rarely gets a hearing in policy debates. That the US Senate saw fit to include this perspective in its first public hearing on drone warfare further illustrates the fleeting but significant breakdown in elite consensus over the controversial program described in Chapter 3. Predictably, corporate media accounts of the hearing foreground divisions within this elite consensus, while downplaying or avoiding altogether al-Muslimi's testimony regarding the drone campaign's toll on civilians. Thus, news coverage of the Senate hearings further illustrates the political struggle inherent in the field of witnessing to the drone wars.

Can I Get a Witness?

Along with eyewitnesses and audiences, mediators—"the various agents and agencies that film, direct, edit, produce, archive, and broadcast testimonials"—are central players in the field of media witnessing (Ashuri & Pinchevski, p. 138). In the case of Farea al-Muslimi's testimony before the Senate, there is considerable slippage between these different players. That is to say, just as al-Muslimi assumes a dual role, as eyewitness to the devastation caused by US drone strikes on one hand, and as a mediator for victims and survivors of these assaults on the other; so too, members of the Senate Judiciary Committee function as mediators who organized the hearings, and as a proximate audience for those testifying. Moreover, the number of agents and agencies involved in mediating the hearings complicate the way audiences access, interpret, and ultimately evaluate al-Muslimi's testimony.

For television audiences, C-SPAN mediates the historic hearings through its live coverage. Subsequent reports mediate this event through a variety of textual, aural, and audiovisual modes. Finally, video clips posted to YouTube and other social media platforms further mediate the hearings, or portions thereof, to audiences removed in both time and space from the live event. Such is the increasingly complex and multifaceted character of witnessing in the era of "spreadable media" (Jenkins, 2009). Here, then, I want to explore how various news organizations construct meaning from al-Muslimi's testimony.

C-SPAN

Prior to the introduction of C-SPAN 2, in 1986, the immediate audience for Senate hearings was limited to a relatively small number of perhaps 100 people who might be in attendance. Through live telecasts, C-SPAN multiplies this audience by an order of magnitude. Today, C-SPAN reaches 95 million American households with live "gavel to gavel" coverage—unadorned by narration or running commentary—of all manner of federal government proceedings. C-SPAN's "unique minimalist model of content packaging" sets it apart from more conventional news channels (Kurpius & Mendelson, 2002, p. 587). Nonetheless, this minimalist approach employs the same resources all mediators do—authorship, narrative, and genre—to construct meaning from raw events. As Ashuri and Pinchevski observe: "In every act of mediating authorship is announced, either directly ... or indirectly" (p. 145). C-SPAN establishes its authorship, and by extension its prestige and credibility, primarily through the use of graphics that identify the network itself, as well as the various speakers at the Senate hearing. That these participants are US Senators, legal experts, and former military officials adds to C-SPAN's reputation as a trusted source for live coverage of important issues and events.

"Through their authorship," Ashuri and Pinchevski remind us, mediators construct narratives of raw events by providing "a timeline, context, circumstance, and causality" (p. 145). At first glance, C-SPAN's minimalist aesthetic seems to preclude the imposition of narrative in rendering the hearings meaningful for television audiences. As we shall see, subsequent news coverage of the Senate proceedings clearly and unambiguously construct a narrative that places al-Muslimi's testimony in a broader context, thereby assigning meaning to his comments vis-à-vis the testimony of others, the questions posed by lawmakers, and the dispute between Congress and the Obama administration over the secretive drone program. In contrast, C-SPAN's live, unscripted, and uninterrupted coverage does not impose a narrative structure on the hearings. Nevertheless, subtle but no less meaningful editorial decisions—such as the aforementioned cutaway to committee members during al-Muslimi's testimony—contribute a sense of narrative tension to the televised proceedings. Likewise, C-SPAN cameras position and frame speakers such that they appear to speak directly to the television audience, thereby inviting a sense of identification, sympathy, and perhaps, empathy in viewers. And while the Senate committee established the order in which various witnesses spoke, as the final witness of the day, al-Muslimi's prepared statement offers a dramatic counterpoint to previous testimony that dealt largely in legal abstractions and technocratic arguments.

Significantly, C-SPAN constructs this narrative through a hybrid genre that combines elements of live reporting with those of documentary. As Ashuri and Pinchevski note: "Mixing genres is yet another resource available to mediators in certain cases" (p. 145). Such is the case with C-SPAN's minimalist approach to public affairs reporting. That is, the hearings unfold in real time—with an intimacy and immediacy that are hallmarks of live broadcasting. By contrast, documentaries operate at a temporal remove from the events being described. Both live reporting and documentary do of course employ testimonials—albeit for different reasons and to different ends. In live reporting, testimony bridges the "time gap between the event and the report on it" (p. 145). Documentaries employ testimony in much the same fashion, but often do so "to add an emotional dimension rather than concrete knowledge" of events that have transpired (p. 145). So it is with al-Muslimi's testimony—most notably the palpable sense of frustration, loss, and grief he conveys—which adds emotional weight to an otherwise dispassionate and bureaucratic exercise in government oversight.

Corporate Media

Headlined "Dems, GOP press Obama administration on drone use," the Associated Press report frames the historic Senate hearings in terms of a breakdown in elite consensus over the drone program. Highlighting the bipartisan nature of the proceedings, the lead paragraph repeats and amplifies this framing. Authored by staff writer Donna Cassata, the story unfolds as a showdown between Congress and the Obama administration over the targeted killing program. Noting that the administration "declined to send a witness to the Judiciary subcommittee meeting," Cassata repeats the White House's longstanding assertion that it derives constitutional authority to use drones for targeted assassination from the 2001 Authorization for Use of Military Force (Cassata, 2013). Thus, Cassata provides context for the hearings *prior* to relating significant details on the proceedings, such as who appeared before the committee and what was said. Placing background information in the lead paragraphs of the report rather uncharacteristically eschews the inverted pyramid style associated with the genre of political journalism.

Midway through her story, Cassata describes a "dramatic moment" when al-Muslimi tells lawmakers that "drones are the face of America" for his people (Cassata). But aside from Retired Marine Corps General John Cartwright, who expressed his concern that "we may have lost the moral high ground" in the war on terror, Cassata does not quote any other panelist. Rather, the story introduces reporting on casualty estimates produced by human rights groups. The piece concludes with a series of quotes from both Democratic and Republican lawmakers who have "created unique alliances" in demanding greater transparency from the Obama administration. Senator Al Franken's remark—"You know you're in strange territory when Senator Cruz and I have the same questions"—supports and reaffirms the story's overarching narrative: a breakdown in elite consensus has produced "odd political bedfellows" (Cassata). In short, the report's structure avoids the inverted pyramid format of political journalism, in favor of a narrative emphasis on rare bipartisan cooperation that epitomizes the genre.

Covering the Senate hearing for the Washington *Post*, Ernesto Londoño (2013) describes the proceedings as "a spirited and unusually public debate." The report's narrative structure reflects this framing—albeit with a degree of ideological closure that serves to legitimate the targeted killing program. Headlined "Drones cause 'growing hatred of America,' bipartisan Senate panel told," Londoño's report bears the hallmarks of "he said, she said" style reportage—an all too common if unproductive approach to political journalism. That is to say, the story unfolds as a series of quotes, first from al-Muslimi and then Senator Durbin. Following this, Londoño inserts an earlier statement by President Obama about his willingness to cooperate with Congress. We then hear from two additional witnesses at the Senate hearing: Rosa Brooks, a Georgetown University law professor, and Retired Air Force Colonel Martha McSally, in that order. The result of this style of political reporting: a litany of competing truth claims devoid of critical evaluation or assessment. While this genre of reporting routinely produces more heat than light, it typifies the political struggle inherent in the field of media witnessing.

Londoño (2013) quotes al-Muslimi in the second paragraph: "What radicals had previously failed to achieve in my village, one drone strike accomplished in an instant: There is now an intense anger and growing hatred of America." Unlike the AP story, then, al-Muslimi gets "top billing" in this account. Nonetheless, Londoño's word choice subverts the meaning and significance of al-Muslimi's testimony. For instance, Londoño avoids mentioning civilian casualties, preferring instead the ambiguous phrase "unintended consequences" to describe the effects of the US drone campaign. Likewise, Londoño characterizes the attack on al-Muslimi's village as a "suspected US drone strike" thereby casting doubt on American culpability for the assault. The balance of the story diverts attention from al-Muslimi's assertions by focusing on the debate between legal, political, and military elites.

Suggesting that terror suspects "may well deserve [their] fate," Professor Brooks nevertheless argues that the secrecy surrounding the drone program "undermines the rule of law" (Londoño). Rather than pursue this line of thinking, Londoño instead quotes McSally, who lauds the technical and legal apparatus behind the drone program in bureaucratic language: "The time between strike approval and weapons release is minimal, maximizing the opportunity to reach the desired effect" (Londoño). This "exchange" exemplifies the "he said, she said" genre of political journalism insofar as it creates a false equivalence between two different perspectives: one that questions the legality of targeted killing, and another that lauds the program's efficiency. The *Post* story ends on a note of pathos based on Mr. al-Muslimi's original reporting. In Yemen, Londoño writes, parents have taken to using drones as a threat to recalcitrant children at bedtime. "'Go to sleep or I will call the planes,' [Al-Muslimi] said, quoting a parental tactic he recently learned about" (Londoño). Notwithstanding the peculiar inversion of Yemeni parents threatening their children with American warplanes—an ideological sleight of hand that deflects blame for the terrifying reality of drone warfare away from the United States– this news story epitomizes the political struggle inherent in media witnessing, and to the ways this contest is constructed discursively.

Like the previous examples, Charlie Savage's report for the New York *Times* bears the imprint of two authors—the news organization and the individual reporter. "Every media report is the product of a media agency which has its own public standing, reputation, and profile" (Ashuri & Pinchevski, p. 145). Like the *AP* and the Washington *Post*, the *Times* is an authoritative source for news from across the nation and around the world. Unlike these other reports, however, Savage's account headlined "Drone Strikes Turn Allies Into Enemies, Yemeni Says," leads with a brief description of Mr. al-Muslimi and his assessment of the drone program before the Judiciary Committee. The next paragraph employs the same pull quote Londoño uses in his piece: "What the violent militants had failed to achieve, one drone strike achieved in an instant" (Savage, 2013). Following this, Savage adds a short paragraph on al-Muslimi's assertion that had they been consulted, local authorities could have captured the individual targeted that day. This portion of al-Muslimi's testimony is conspicuously absent from both the *AP* and *Post* stories. However, Savage only summarizes this portion of al-Muslimi's testimony without benefit of direct quotation.

Savage continues that while some (unnamed) legal experts at the hearing called for greater transparency regarding the standards and procedures used in selecting individuals for targeted assassination, others suggested that the weapons employed in these operations are "no different from other means of lethal force" (Savage). Again, Savage avoids direct quotation, preferring instead to assert his authorial voice in this account. Significantly, Savage gives Colonel McSally the final word—albeit by paraphrasing McSally's testimony. In the penultimate paragraph, he writes:

> She also said that even when it appeared possible to arrest someone rather than killing him, there were risks of civilian casualties in "capture missions," too. Such operations also take longer to plan, and create the risk that an American commando might be killed or captured, she said. (Savage)

Ostensibly a story about the strategic failure of the drone program, Savage concludes his report with McSally's resounding defense of targeted killing. Doing so, Savage echoes and amplifies well-worn Pentagon and White House talking points that justify and normalize targeted killing. Equally important for my purposes here, the *Times*' report further illustrates the contested character of media witnessing—and the decisive role mediating agents and agencies play in this process. Put differently, ending his report with McSally—an unequivocal advocate of the drone program—Savage all but eclipses al-Muslimi's damning testimony.

Public Media

US public broadcasting's sparse coverage of the Senate's first public hearing on drone warfare reflects two aspects of corporate media's reporting: near silence on the one hand, and cool reassurance on the other. Like the majority of its network television counterparts, PBS failed to cover the Judiciary Committee's proceedings. Indeed, apart from C-SPAN's coverage and a single report on MSNBC, the liberal-leaning cable news network, there was a virtual blackout of the Senate hearings on American television. For its part, NPR's coverage echoed corporate media narratives that highlight the breakdown in elite consensus over the US drone program, while studiously avoiding the question of civilian casualties. Titled "Lawmakers Want To Avoid Drone Strike Abuses," the archived report and online transcript reveal NPR's deference to American political and military leadership in the matter of targeted killing.

In their introduction to correspondent Carrie Johnson's report, *Morning Edition* co-hosts Steve Inskeep and David Green frame the story in terms of the secrecy surrounding the targeted killing program and Congressional demands for greater transparency.

INSKEEP:	Until very recently, this next story was being discussed in public by seemingly everyone except the United States government.
GREENE:	It's the US policy of targeted killing, commonly using drones buzzing overhead to strike suspected militants in places like Pakistan or Yemen. President Obama's administration has only begun to discuss the program.
INSKEEP:	And yesterday the Senate held its first public hearing. Lawmakers say they want to establish some rules and prevent abuse. NPR's Carrie Johnson reports. (Johnson, 2013)

Like the newspaper and wire service stories discussed above, a byline signals the authorship of NPR's report. Unlike print, however, radio journalists also claim authorship through their vocal presence in the stories they organize and narrate. To that end, the body of the report consists of two discrete production elements: Johnson's narration (tracks) and actualities (acts) of various participants at the Senate hearing.[3]

Johnson opens her report noting the White House's refusal to send a representative to meet with lawmakers. "That's something one of the President's closest allies, Senator Richard Durbin of Illinois, pointed to right out of the gate" (Johnson). Like corporate media accounts, NPR's report emphasizes the breakdown in elite consensus over the President's authority to use drones for targeted killing. Interspersing her narration with sound bites of Senator Durbin's opening remarks, Johnson lays out the constitutional questions underlying the Senate proceedings.

SENATOR DURBIN:	I do want to note for the record my disappointment that the administration declined to provide a witness to testify at today's hearings.
JOHNSON:	Durbin, a Democrat, said the law hasn't kept up with technology or the facts on the ground.
SENATOR DURBIN:	I knew what we were voting for in 2001 when it came to Afghanistan. That's where al-Qaida was. And they had just attacked the United States and we were going to answer that attack. I didn't realize that we would be having this conversation 12 years later about Yemen, Somalia, even Pakistan. (Johnson)

Durbin's comments reflect growing unease with the Obama administration's escalation and expansion of the drone program across the Greater Middle East. Likewise, Georgetown University law professor Rosa Brooks expresses her apprehension over the constitutional implications of the targeted killing program. Here, then, Johnson identifies the secrecy surrounding the program as the crux of the matter.

BROOKS:	Right now we have the executive branch making a claim that it has the right to kill anyone anywhere on Earth at any time for secret reasons based on secret evidence in a secret process undertaken by unidentified officials. That frightens me. (Johnson)

Brooks' comments underscore the constitutional implications of unchecked executive authority. Left largely unsaid, however, in this and other portions of Senate hearings, are the human costs and consequences of the targeted killing program. Indeed, throughout much of the hearing, panelists frequently used sanitized language to elide and obscure the issue of civilian casualties in America's drone wars. Rather than interrogate this evasive language, Johnson's report echoes and amplifies these prevarications.

JOHNSON:	Brooks said there may well be a sound legal reason for all of those US drone strikes. But people just don't know enough to feel comfortable about it. Retired General James Cartwright.
GENERAL CARTWRIGHT:	I am concerned we may have ceded some of our moral high ground in this endeavor.

JOHNSON:	Cartwright is the former vice chairman of the Joint Chiefs of Staff, almost as high as a person can reach on the military ladder.
GENERAL CARTWRIGHT:	In several areas around the world, current drone policies have left us in a position where we are engendering more problems than we're solving. (Johnson)

At first blush, Cartwright's "concerns" suggest principled opposition to the drone program. However, this muted critique fails to forthrightly acknowledge the persistent moral and ethical questions raised by this new kind of war. In Johnson's words, critics don't "feel comfortable" with the program's secrecy—less so the devastating impact of drone warfare on innocent civilians, it would seem. Likewise, Cartwright asserts that, despite these unmentioned and seemingly unmentionable civilian casualties, the United States "may" have lost the moral high ground. As a result, Cartwright concludes, the program creates "more problems" for the United States. With an unmistakable air of American exceptionalism, Cartwright's criticism, like Johnson's reporting, neglects the carnage unleashed by the targeted killing program and focuses instead on vague and ambiguous "abuses."

At this juncture, Johnson introduces Farea al-Muslimi and briefly summarizes his testimony. The single actuality of al-Muslimi's statement is the same pull quote used in the aforementioned Washington *Post* and New York *Times* stories. Dramatic to be sure, this sound bite overshadows much of al-Muslimi's testimony, which bore witness to the civilian toll of US drone strikes; thereby reducing his testimony to an assessment of the efficacy of US counterterrorism strategy in winning hearts and minds, while studiously avoiding the broader civilian impact of the American drone campaign across Yemen.

NPR's story is one of the few domestic news accounts to even mention the testimony of Peter Bergen: a vocal critic of the drone program's toll on civilians. Nevertheless, Johnson's report sets up Bergen as a foil to Colonel McSally's spirited defense of targeted killing. Reminiscent of the "he said, she said" style of political journalism described above, Johnson's narrative brings Bergen and McSally into dialogue. Unlike McSally, however, we do not hear from Bergen in his own words. And as it happens, Johnson's report uses the same comforting sound bite from McSally's testimony referenced by both the *Post* and the *Times*.

Johnson concludes that Senator Lindsey Graham, a frequent critic of the President, "applauded the Obama administration for its aggressive and responsible approach to drones. Many of his colleagues remain uneasy," Johnson adds. "Senator Durbin says he'll have more hearings later this year." NPR's report concludes in a decidedly upbeat fashion; one designed to mitigate the breakdown in elite consensus and reassure listeners that lawmakers will do all they can to curtail "abuses"—a euphemism for civilian casualties. Put differently, NPR's story offers listeners cool reassurance, not only of the efficiency of targeted killing, but also for the prospects of greater Congressional oversight of the drone program. Like corporate media coverage of al-Muslimi's testimony, NPR's account of the Senate hearings glosses over the issue of civilian casualties, focusing instead on a short-lived breakdown in elite consensus and ultimately lending legitimacy to this new kind of war.

In contrast to the highly circumscribed coverage of al-Muslimi's moving testimony, several news outlets—the cable network MSNBC, the online magazine Salon.com, and Pacifica radio's *Democracy Now!*—supplemented their reports on the Senate hearings with follow-up interviews with al-Muslimi. For instance, when asked about a May 2011 drone strike that targeted a suspected terrorist, Anwar al-Arshani, and "inadvertently" killed dozens of civilians, al-Muslimi told Salon:

> Al-Arshani's house was the first drone strike I ever visited. I interviewed most of the survivors. I visited them almost a month after the strike, and I still found a few pieces of his house. It is one of the places in Yemen where most civilians were killed ever [as a result of a drone strike]. Whether al-Arshani was part of al-Qaida or not, the US killed 24 innocent Yemenis. (Ali, 2013)

An altogether common journalistic practice, it is all the more revealing how few news organizations saw fit to conduct interviews with such an articulate, personable, and knowledgeable witness. NPR's failure to invite al-Muslimi to discuss his reporting is especially egregious in light of his work as a translator for the public broadcaster during the Arab Spring.[4] Without putting too fine a point on it, then, the virtual blackout of al-Muslimi's dramatic testimony, coupled with the failure of most news organizations to interview the young reporter, reveals just how difficult it is for alternative perspectives on the targeted killing program to enter mainstream news discourse. Nevertheless, as I have suggested here and in the preceding chapter, witnesses are eager to tell their stories. Equally important, the fact that al-Muslimi's testimony went viral suggests that despite, or more accurately perhaps, because of the media silence surrounding civilian casualties in the drone wars, significant portions of the American public are just as eager to hear their stories.

All of which begs several crucial questions. Having heard these stories, are audiences obliged to do something about it? Put differently, what are the responsibilities of contemporary media audiences when confronted with the pain and suffering of distant others? Are we, as witnesses to mediated atrocity, or as an audience for such witnesses, somehow complicit in this suffering? After all, in an era when images and accounts of atrocity circulate worldwide "we cannot say that we do not know" (Ellis, 2000, p. 1). This fundamental problem motivates recent thinking on the relationship between media witnesses and their audiences. A relationship made all the more significant in light of the affordances of social media for watching, sharing, and responding (individually or collectively) to distant suffering. In short, contemporary audiences "are not simply observers, they are, at least potentially, judge and jury" (Ashuri & Pinchevski, p. 140).

It is precisely this potential—for audiences to witness such testimony, judge it for themselves, and then formulate a response—that motivates people like Farea al-Muslimi to enlist the American public's support in putting an end to killing by remote control. "I deeply believe," al-Muslimi told lawmakers, "that when Americans truly know about how much pain and suffering these airstrikes have caused … they will reject this devastating targeted killing program" (C-SPAN). Rafiq Rehman, the Pakistani school teacher whom we met in the previous chapter, expressed this very same sentiment when he told reporters, "That's why I came here—I thought if they heard my story, they would want to listen to me and influence their politicians" (Khan, 2013). As we shall see in the next chapter, this same realization motivates drone whistleblowers who likewise bear witness to the brutality of robotic warfare, and challenge audiences to judge what they see, hear, and read. All of which underscores the unequal power/knowledge relationships in the field of media witnessing, and the decisive role mediators (journalists and news organizations) play in relaying, or more frequently, marginalizing the stories and perspectives of victims, survivors, and witnesses to America's drone wars.

Notes

1. C-SPAN 3 carried the Senate hearings live. C-SPAN 2 re-broadcast the proceedings later that day. The entire hearing is available via the C-SPAN archives. See https://www.c-span.org/video/?312317-1/senate-committee-examines-legality-us-drone-strikes.
2. In graphic detail, al-Muslimi also describes a December 17, 2009 cruise missile attack on the village of al-Majalah that killed upwards of 40 civilians—including pregnant women and children. My focus on drone warfare elides other equally pernicious military tactics such as night raids, firefights, and air assaults conducted with guided missiles and manned aircraft.

3. Here I use NPR's terms to describe production elements in a radio news story. See "What Does a Radio Script Look Like?" *NPR*, http://training.npr.org/audio/what-does-a-radio-script-look-like/.
4. See, for example, this report from June 2012 that describes the Yemeni central government's dismantling of a protest camp in the capital city Sana'a. "Is a protest camp still needed in Yemen?" *NPR*, http://www.npr.org/2012/06/27/155866875/is-a-protest-camp-still-needed-in-yemen.

References

Ali, W. (2013, May 2). Drone victim: US strikes boost al-Qaida recruitment. *Salon*. Retrieved from http://www.salon.com/2013/05/02/drone_victim_u_s_strikes_boost_al_qaeda_recruitment/

Al-Muslimi, F. (2013, April 23). Statement of Farea al-Muslimi. *Drone wars: The constitutional and counterterrorism implications of targeted killing*. Submitted to the Senate Committee on the Judiciary, Subcommittee on the Constitution, Civil Rights and Human Rights. Retrieved from https://www.judiciary.senate.gov/imo/media/doc/04-23-13Al-MuslimiTestimony.pdf

Ashuri, T., & Pinchevski, A. (2011). Witnessing as a field. In P. Frosh & A. Pinchevski (Eds.), *Media witnessing: Testimony in the age of mass communication* (pp. 133–157). New York, NY: Palgrave Macmillan.

Cassata, D. (2013, April 23). Dems, GOP press Obama administration on drone use. *Associated Press*. Retrieved from https://www.yahoo.com/news/dems-gop-press-obama-administration-drone-205450106--politics.html?ref=gs

C-SPAN. (2013, April 23). *Targeted killing of terrorist suspects overseas*. Retrieved from https://www.c-span.org/video/?312317-1/senate-committee-examines-legality-us-drone-strikes

Durbin, R. (2013, April 23). Opening statement. *Drone wars: The constitutional and counterterrorism implications of targeted killing*. Submitted to the Senate Committee on the Judiciary, Subcommittee on the Constitution, Civil Rights and Human Rights. Retrieved from https://www.judiciary.senate.gov/imo/media/doc/04-23-13DurbinStatement.pdf

Ellis, J. (2000). *Seeing things: Television in the age of uncertainty*. London: I. B. Tauris.

Frosh, P., & Pinchevski, A. (Eds.). (2011). *Media witnessing: Testimony in the age of mass communication*. New York, NY: Palgrave Macmillan.

Jenkins, H. (2009). *If it doesn't spread, it's dead (part one): Media viruses and memes*. Retrieved from http://henryjenkins.org/2009/02/if_it_doesnt_spread_its_dead_p.html

Johnson, A. (2013, April 24). Lawmakers want to avoid drone strike abuses. *NPR*. Retrieved from http://www.npr.org/2013/04/24/178753575/lawmakers-want-to-avoid-drone-strike-abuses

Katz, E. (2011). Preface to the paperback edition: No more excuses. In P. Frosh & A. Pinchevski (Eds.), *Media witnessing: Testimony in the age of mass communication* (pp. vii–viii). New York, NY: Palgrave Macmillan.

Khan, N. (2013, October 29). Drone victims give firsthand account of attack. *Al Jazeera America.* Retrieved from http://america.aljazeera.com/articles/2013/10/29/congressional-briefingonpakistanidronevictimsdraws5lawmakers.html

Kurpius, D., & Mendelson, A. (2002). A case study of deliberative democracy on television: Civic dialogue on C-SPAN call in shows. *Journalism & Mass Communication Quarterly, 79*(3), 587–601.

Londoño, E. (2013, April 23). Drones cause "growing hatred of America," bipartisan Senate panel told. *Washington Post.* Retrieved from https://www.washingtonpost.com/world/national-security/drones-cause-growing-hatred-of-america-bipartisan-senate-panel-told/2013/04/23/4863b1f8-ac6e-11e2-a8b9-2a63d75b5459_story.html

Nye, J. S., Jr. (1990). Soft power. *Foreign Policy, 80*, 152–171.

Rentschler, C. (2011). From danger to trauma: Affective labor and the journalistic discourse of witnessing. In P. Frosh & A. Pinchevski (Eds.), *Media witnessing: Testimony in the age of mass communication* (pp. 158–181). New York, NY: Palgrave Macmillan.

Savage, C. (2013, April 23). Judiciary panel hears testimony on use of drones. *New York Times.* Retrieved from http://www.nytimes.com/2013/04/24/world/middleeast/judiciary-panel-hears-testimony-on-use-of-drones.html?_r=0

· 9 ·

DISTRIBUTED INTIMACIES

Robotic Warfare and Drone Whistleblowers

Technology is making gestures precise and brutal, and with them men.
—Theodor Adorno, *Minima Moralia: Reflections from Damaged Life*

In May 2009, *60 Minutes*, the long-running CBS News magazine, published a report by correspondent Lara Logan profiling the US drone program.[1] Despite the classified nature of these operations, Logan and her crew were granted "secret clearance and unprecedented access" to Creech Air Force base outside Las Vegas, "the first base in Air Force history that exclusively flies unmanned aircraft." The result: an unambiguously upbeat assessment of what the news magazine called "America's new Air Force." Whether there was a quid pro quo involved in granting the *60 Minutes* crew access to Creech remains unclear. Nevertheless, Logan's praise for the new weapon system that "fundamentally rewrites the rules of battle" bears the unmistakable imprint of the technological sublime. During a brief but revealing exchange between Logan and base commander, Colonel Chris Chambliss, the veteran F-16 fighter pilot dismisses critics' claims that, unlike their counterparts flying manned aircraft in the skies above Afghanistan and Iraq, drone pilots are emotionally and psychologically detached from combat. "I heard those arguments," Chambliss assures the reporter. "And I can tell you that—and I'm happy to tell ya—that I've never been more engaged in conflict in my life." "And," Logan adds helpfully, "he's never been safer."

Herein lies the cruel irony of drone warfare. On the one hand, digital communication and surveillance technologies enable drone operators[2] to experience combat with unprecedented intimacy; an intimacy not only with the ground troops and special operations forces they assist and support, but also with the suspected militant groups and individuals they track and kill. On the other hand, thanks to these same globe-spanning technologies, drone operators do so from half a world away—well out of harm's way. Hence, the sublime appeal of robotic warfare: operating from the safety and security of ground control stations (GCS) located thousands of miles from the battlefield, remotely piloted aircraft (RPA) spare US military personnel from the "dull, dirty, and dangerous" realities of combat (Irvine, 2012). Or so it would seem. For as I aim to demonstrate throughout this chapter, the scale, scope, and reach of this intimacy is unprecedented in the annals of warfare, and carries with it enormous burdens and responsibilities. Furthermore, what I'm calling "distributed intimacy"—asymmetrical relations of power and knowledge between America's global armed forces and the tribal communities they surveil and assault—raises profound legal, strategic, and ethical questions that deeply trouble some drone operators. Effectively turning these remote-control warriors into digital witnesses of atrocity.

Researchers characterize the socio-technical nexus between digital media and intimate lives in terms of "digital intimacies" (Sadowski, 2016). As digital devices infiltrate public and private space, and develop increasingly sophisticated means of recording, analyzing, and distributing personal information and behaviors, recent scholarship has explored the shifting nature of intimate relationships in electronic space. Typically, researchers conduct their analyses through the lens of promotional and brand culture (Banet-Weiser, 2012; Carah, 2017; Winch, 2015). This line of inquiry underscores the commodification of intimate and everyday lives through all manner of calculating devices and platforms: from mobile phones and laptop computers to social media and dating apps. However, this commercial logic is but one tendency of digital mediation. The emergence of drone strikes as a favored tactic in America's war on terror reveals a far more ominous tendency: the authoritarian logic of calculative media. That is to say, just as brand culture constructs and promotes entanglements between lived experience and digital media for purposes of marketing and profit making, so too military and paramilitary organizations leverage calculative media and "pattern-of-life analysis" (Franz, 2017) to conduct targeted assassination and robotic warfare. Put differently, the use of unmanned planes to locate, track, and kill suspected terrorists *weaponizes* digital intimacies.

This chapter adapts the concept of "distributed intimacy" (Cohen, 2010; Edmondson, 2001) in an effort to identify and analyze unequal relations of power/knowledge in the mediated relationships articulated by this new kind of war. Throughout, I contend this notion enhances our understanding of the authoritarian logic of disembodied control at a distance underpinning America's drone wars. The chapter proceeds in three parts. The first develops an analytical framework for examining the distributed intimacies engendered and exploited by drone warfare. Doing so, I identify revealing points of comparison between commercial and authoritarian logics of digital mediation. The second considers the affective and political consequences of this new kind of war for drone operators turned whistleblowers. Here I consider the relationship between digital witnessing and trauma (Chouliaraki, 2015; Rentschler, 2011) in the era of robotic warfare. Based on an examination of press accounts, broadcast interviews, and documentary films, this chapter identifies drone whistleblowers—whose intimate testimony exposes the physical, emotional, and psychological brutality of drone warfare—as central actors in the formation of an alternative order of discourse surrounding weaponized drones. The chapter concludes with an assessment of the personal and institutional challenges confronting the ranks of remote-control warriors as Donald Trump, one of the most authoritarian figures in recent American history, assumes the office of the presidency, and with it, command and control of the US drone program.

Distributed Intimacies

In his analysis of drone warfare and its implications for just war theory, John Williams (2015) introduces the concept of "distant intimacy" to describe the uncanny spatial relationship between drone operators and their quarry engendered by this new kind of war. For Brandon Bryant, a former USAF Staff Sergeant turned whistleblower, the perverse nature of this intimacy is clear.

> We are the ultimate voyeurs, the ultimate peeping Toms. I'm watching this person, and this person has no clue what's going on. No one's going to catch us. And we're getting orders to take these people's lives. It was just point and click. (Løge & Schei, 2014)

Of course, drones are not unique in their capacity to exploit the strategic advantage of waging war at a distance. Long before David slew Goliath, the dream of striking an opponent from a safe distance has been manifest in all manner of human expression. From ancient mythology and science fiction to technological research and development, distancing weapons have been and remain an object of fascination and desire. Such is the sublime appeal of conducting war from ever-greater distances—and with ever-greater impunity—described in Chapter 1.

For present purposes, I draw upon Williams's observations on the visual dimensions of the distance intimacy articulated through drone warfare. Given the aircraft's ability to closely monitor and observe suspected militants for hours and days at a time, Williams contends remote operators learn a great deal about these individuals—their private lives, personal behaviors, and social contacts; far more intimate knowledge than anything produced by conventional weapon systems. Indeed, the novelty of robotic warfare's distant intimacy is frequently celebrated in news and popular culture. Consider, for example, *The Huffington Post's* description of veteran drone pilot Bill "Sweet" Tart's unique aerial perspective:

> He watches for the kids. Peering through cameras and sensors from his computer station thousands of miles away, he absorbs the details of daily life in the villages below. He develops an eerie intimacy with his targets. Which house these kids belong to. When that mom goes out to market. Who visits and why. He tries to ensure innocents are nowhere near. (Wood, 2013)

Notwithstanding this story's reassuring tone, the distance intimacy between drone operators and their targets reveals a great deal about the everyday lives and social interactions of individuals, groups and entire villages. Speaking with *Foreign Policy* magazine, an unnamed remote pilot describes this relationship in clinical and ethnographic terms.

> Flying an RPA, you start to understand people in other countries based on their day-to-day patterns of life. A person wakes up, they do this, they greet their friends this way, etc. You become immersed in their life. You feel like you're a part of what they are doing every single day. So, even if you're not emotionally engaged with those individuals, you become a little bit attached. I've learned about Afghan culture this way. You see their interactions. You're studying them. You see everything. (Rothenberg, 2014)

Conversely, in their role as force protectors, drone operators share unprecedented intimacy with US ground troops in Afghanistan, Iraq and elsewhere

(Gregory, 2011). Describing the camaraderie and emotional investment that comes with flying a drone for close air support, rather than a manned vehicle, the same remote pilot continued:

> Because of the length of time that you're over any certain area you're able to engage in lengthy communications with individuals on the ground. You build relationships. Things are more personal in an RPA than an aircraft that's up for just a few hours. … And with that, there's an emotional attachment to those individuals. (Rothenberg)

In short, the drone's advanced optical, communication, and data transmission systems restore a level of intimacy to modern warfare long since displaced by the advent of distancing weapons such as fighter jets and ballistic missiles. Nevertheless, as Williams contends, the distance intimacy afforded by drone warfare is incomplete and unambiguously one-sided: "this partial restoration of intimacy is asymmetric: only the drone operator gains insight into the life of his adversary, sees his face, and witnesses his pain and death" (p. 96). For whole communities living under drones, no such intimacy exists: only the dread realization that unmanned surveillance vehicles carry lethal payloads that can be deployed in an instant. Thus, the asymmetrical intimacy produced by these rapid-reaction, force projection technologies expresses and reinscribes power differentials between drone operators and their targets.

Visual surveillance is but one dimension of the distant intimacies engendered by robotic warfare. In April 2014, General Michael Hayden, who served as director of both the NSA and CIA, publicly acknowledged for the first time the role calculative media and pattern-of-life analysis (POL) play in conducting US drone strikes. Speaking at a foreign affairs symposium at Johns Hopkins University, Hayden put it plainly: "We kill people based on metadata" (Cole, 2014). Notwithstanding the chilling privacy implications of all this, Hayden's admission confirms the existence of a semi-autonomous process for selecting targets based not on a known individual, but rather a potential target's suspicious patterns of behavior or "signature"; hence the distinction between so-called personality and signature strikes respectively (Williams, 2013). As Nina Franz (2017) concludes: "In addition to video surveillance, this signature may consist of electronic communications that the CIA is known to collect through technology installed on drones via the NSA, allowing the US intelligence agencies to gather data indiscriminately 'on an industrial scale'" (p. 116).

Here we can detect a correspondence between the commercial and authoritarian logics of calculative media. That is to say, just as corporations from Amazon to Wal-Mart "harness, mine and infiltrate social networks and private relationships" (Winch, 2015, p. 229) to deliver targeted marketing and advertising content to consumers, so too do US intelligence agencies, military and paramilitary forces analyze "digital footprints" (Weaver & Gahegan, 2007) gleaned from cell phones, laptops, and computer servers to target groups and individuals for summary execution. And, like brand culture that employs calculative media to "penetrate the intimate connections" (Winch, p. 238) of consumers, robotic warfare collects, analyzes, and stores metadata from digital devices to profile suspected militants based on age, gender, location, and personal affiliations. This so-called signals intelligence (SIGINT) provides the basis for subsequent "actionable intelligence": a military euphemism for the expedited use of deadly force. Working alongside the CIA and the Joint Special Operations Command (JSOC) to conduct drone strikes, the NSA's geolocation unit adopted a motto that succinctly captures the collaboration between US intelligence agencies and armed forces: "We Track 'Em, You Whack 'Em" (Priest, 2013). Whereas the commercial logic of calculative media aims to monetize digital intimacies, the authoritarian logic weaponizes these intimacies.

Writing in *The Guardian*, John Naughton likens one such targeting technique, the NSA's SKYNET protocol—a darkly ironic reference to the apocalyptic killing machine of the *Terminator* films—to a spam filter on an email account.

> The critical difference is that if your filter gets it wrong, then the worst that can happen is that you are annoyed or amused by its clumsiness; if SKYNET gets it wrong you could find yourself on the receiving end of a Hellfire missile dispatched by a Predator or a Reaper drone. (2016)

In many respects, then, SKYNET's metadata analysis and algorithmic calculation "works like a typical modern Big Data business application" (Grothoff & Porup, 2016). Still, given SKYNET's scope—the program is said to use a machine-learning algorithm to analyze the metadata of some 55 million Pakistani mobile phone users—the potential for lethal mistakes is considerable. Patrick Ball, director of research at Human Rights Data Analysis Group, characterized NSA's methods as "wildly optimistic" and "complete bullshit" (Grothoff & Porup). Notwithstanding Ball's colorful language, published reports indicate that between 2004–2015, upwards of 2500 people, including as many as 424 to 966 civilians, have been killed by American drone strikes in Pakistan alone (Serle, 2016).

The correspondence between the commercial and authoritarian logics of calculative media does not end there. "The relationship between marketer and consumer requires a media infrastructure that can monitor actors and modulate their action over time" (Brodmerkel & Carah, 2016, p. 145). So too, the US drone program relies on a global network infrastructure, the Distributed Common Ground System (DCGS), to gather, analyze, and act upon the voluminous data used to conduct targeted killing operations (USAF, 2015). I discuss the theoretical implications of this global network in the book's final chapter. For now, suffice it to say the DCGS distributes this torrent of data far and wide: from remote pilots and ground troops to teams of analysts and supervisors, up to and including high ranking military and political leaders. In other words, the digital intimacies generated and exploited by an increasingly expansive global surveillance regime are distributed up and down what the military calls "the kill chain" (Cockburn, 2015).

Contrary to depictions in news accounts and popular entertainment, robotic warfare involves more than a handful of remote pilots, intelligence analysts, and their superiors. Rather, a sprawling clandestine bureaucracy staffed by military and paramilitary personnel, and articulated through satellite data links, real-time video feeds, email chat and other forms of networked communication conducts US drone strikes (Asaro, 2013). Speaking with reporters, whistleblower Brandon Bryant neatly conveyed the decentralized and geographically dispersed character of the targeted killing program: "You got to understand that the whole operation procedure is like a web, and, like, you're dealing with people from multiple locations all over the world" (Goodman & Gonzalez, 2013).

Significantly, NSA's decisive role in the targeted killing program first came to light in the classified documents leaked by former CIA contractor Edward Snowden (Mazzetti & Schmidt, 2013). Speaking with filmmaker Laura Poitras and journalist Glenn Greenwald in the Academy Award-winning documentary *Citizenfour*, Snowden revealed that government contractors with security clearance have real-time access to hundreds of video streams acquired through an expansive drone surveillance network. Snowden told filmmakers, "As I saw that, that really hardened me to action" (Soderbergh & Poitras, 2014). The close and previously undisclosed collaboration between US intelligence services, military, and paramilitary forces ultimately led Snowden to disclose classified information about the NSA's unprecedented and unlawful surveillance regime. Likewise, a handful of drone operators—motivated by concerns over the targeted killing program's lack of transparency, its potential for fueling Islamic extremism, and the trauma associated with digital witnessing in robotic warfare—have summoned the courage to expose the tactical, strategic, and ethical fallacies underpinning the US drone campaign. Their witnessing serves to destabilize the dominant order of discourse that celebrates the precision, accuracy, and efficiency of drone warfare.

From Digital Witness to Drone Whistleblower

Unlike sailors who fire submarine-launched cruise missiles (SLCM) from across distant oceans, or fighter pilots who unleash laser-guided bombs from 30,000 feet in the air, drone operators view their deadly assaults in real-time and graphic detail. "We're not disconnected from what's happening. We're not playing videogames. With RPAs, you grasp your enormous level of responsibility. You witness it all. … It is war at a very intimate level" (Rothenberg). In their dual role as perpetrators and observers of targeted killing, drone operators complicate our understanding of "digital witnessing"—what Lilie Chouliaraki defines as "the moral engagement with distant suffering through mobile media" (2015, p. 1362). Chouliaraki examines the prevalence in Western media of amateur recordings of death scenes: blanket news coverage of the capture and killing of Libyan strongman Muammar Gaddafi, jihadist execution videos, and images of civilian casualties in Syria. To that end, her analysis focuses on the active participation of local actors in the unfolding violence and its attendant visual documentation. Chouliaraki concludes: "Digital witnessing is, in this sense, defined by the new status of the camera not only as tool for the professional reporting of conflict, but, simultaneously, as a weapon in the very conduct of conflict" (p. 1363). The sophisticated optical systems required to hunt and kill suspected terrorists likewise redefines the camera as a weapon of robotic warfare.

Nevertheless, there are important distinctions between the footage produced by militant groups and that produced by US forces. First and foremost, local actors are proximate to the events they record, while drone operators conduct their missions from half a world away. And unlike amateur recordings, literally produced on the ground, the aerial perspective engendered by drone warfare further distances the perpetrator from the unfolding violence. Put differently, remote crews are present in time but removed in space from the violence they both inflict and observe. Second, the drone operator's globe-spanning perspective is an exemplar of what Donna Haraway (1997) calls "modest witnessing." That is to say, drone surveillance enacts a scientific mode of observation, marked by (radical) detachment on the one hand, and (unsolicited) intimate inspection on the other. In contrast, amateur recordings from conflict zones offer no such detachment. The currency of this footage, if not its authenticity, rests upon the graphic nature of the violence it depicts—by its very immodesty. Third, the clandestine nature of the drone program, coupled with American news organizations' self-censorship, routinely precludes dissemination of graphic footage from these strikes in US news coverage (Gregory, 2011; Rowley, 2014): a sharp contrast to the incessant replay of jihadi footage of atrocity common to rolling news channels.

And yet, just as the Islamic State publishes execution videos on social media for publicity and recruitment campaigns, the US Department of Defense posts drone strike footage to YouTube for propaganda purposes (Stahl, 2013). Ostensibly meant to facilitate the domestication of drones in civilian airspace by celebrating American technological ingenuity and superiority, the footage further aims to intimidate and demoralize America's enemies (Benjamin, 2013). Nevertheless, "these graphic images rarely make their way to mainstream US news media because such sources typically forgo explicit visuals in favor of more clinical visuals of command centers or drone pilots" (Scharrer & Blackburn, 2015, p. 800). That US news organizations fetishize amateur footage documenting Islamic brutality while studiously avoiding images of American military violence further underscores the ideological dimensions of the deferential and highly circumscribed news coverage of America's drone wars discussed in Chapter 6.

Herein we can discern the discursive and political significance of whistleblowers bearing witness to the brutality of drone strikes: a cruel and cowardly[3] tactic advocates of targeted killing rarely acknowledge. For instance, in a first-person column for *The Guardian*, subsequently republished in *AlterNet*, Heather Linebaugh, a former image analyst, expresses her frustration with lawmakers who laud the precision and accuracy of drone technology.

> Few of these politicians who so brazenly proclaim the benefits of drones have a real clue of what actually goes on. I, on the other hand, have seen these awful sights first hand. I knew the names of some of the young soldiers I saw bleed to death on the side of a road. I watched dozens of military-aged males die in Afghanistan, in empty fields, along riversides, and some right outside the compound where their family was waiting for them to return home from the mosque. (Linebaugh, 2013)

Following Tait's (2011) explication of "bearing witness," Linebaugh's testimony reveals three elements of this discursive practice. First and foremost, bearing witness is fraught with the difficulty of representing "trauma and atrocity through words and images" (p. 1222). Despite the surplus of surveillance footage collected before, during, and after a US drone strike (Gregory), whistleblowers rely primarily on language to convey the brutality of drone warfare.

> Horrifying barely covers it. And when you are exposed to it over and over again it becomes like a small video, embedded in your head, forever on repeat, causing psychological pain and suffering that many people will hopefully never experience. UAV troops are victim to not only the haunting memories of this work that they carry with them, but also the guilt of always being a little unsure of how accurate their confirmations of weapons or identification of hostile individuals are. (Linebaugh)

While drone operators may circumvent the physical dangers of combat, the psychological and emotional harm produced by this new kind of war is manifest in their moving and sometimes graphic testimony. Linebaugh's statement not only reveals the difficulty associated with bearing witness to robotic war, and the hidden toll it takes on remote-control warriors, but also challenges the dominant narrative that promotes the accuracy and precision of America's targeted killing program (Di Salvo, 2014).

Second, Linebaugh's contempt for the willful ignorance of policy makers underscores the affective dimension of bearing witness: "forms of embodied practice that exceed normative renderings of impartiality and detachment" (Tait, p. 1222). Contrasted with the reassuring tones and bureaucratic language of official sources, Linebaugh's indignation is clear from the outset. Her column opens with a series of penetrating questions for supporters of the drone program.

> I wish I could ask them a few questions. I'd start with: "How many women and children have you seen incinerated by a Hellfire missile?"
>
> And: "How many men have you seen crawl across a field, trying to make it to the nearest compound for help while bleeding out from severed legs?"
>
> Or even more pointedly: "How many soldiers have you seen die on the side of a road in Afghanistan because our ever-so-accurate UAVs were unable to detect an IED that awaited their convoy?" (Linebaugh)

Here, the former image analyst leverages her status as an expert and insider, one who has witnessed repeatedly the brutality of robotic war. Significantly, Linebaugh's rhetorical strategy resists professional norms of objectivity and emotional detachment—precisely the sort of modest witnessing associated with scientific, military, and journalistic practice. Rather, her intimate testimony directly contradicts news and official discourse with a highly subjective, morally outraged rendering of the consequences of drone warfare for combatants and civilians alike.

Doing so, Linebaugh enacts the third dimension of bearing witness: the desire to illicit a response from her readers: "a *public reaction* that manifests as *action*, rather than a collective shrug" (Tait, p. 1222). To this end, Linebaugh decries the secrecy surrounding targeted killing, as well as the dubious claims regarding the program's utility as an instrument of counterterrorism. Instead, Linebaugh demands greater transparency and public accountability in an effort to reveal the increasingly prominent role drones play in the growth and expansion of twenty-first century American empire—a topic taken up at length in the final chapter. Finally, Linebaugh's testimony sounds a cautionary note for Americans who have been lulled into a false sense of security: "The UAVs in the Middle East are used as a weapon, not as protection, and as long as our public remains ignorant of this, this serious threat to the sanctity of human life—at home and abroad—will continue" (Linebaugh).

This note of caution runs through a great deal of whistleblower testimony. Brandon Bryant sums up the policy implications of drone warfare this way: "It can't be a small group of people deciding how they're used. There's got to be transparency. People have to know how they're being used so they're used responsibly" (Power, 2013). To that end, Bryant and his colleagues contend the expansion of the targeted killing program under the Obama administration undermines national security by inciting, rather than suppressing, Islamic extremism. In contrast to politicians and pundits who raise the specter of "blowback"—in intelligence parlance the unanticipated and unintended consequences of covert action—to justify further military intervention across the Muslim world, whistleblowers openly address the strategic failure of the US drone campaign.

For instance, in the aftermath of the November 13, 2015 terrorist attacks in Paris, a cadre of whistleblowers took the unprecedented step of publishing an open letter to President Obama, calling for an end to the targeted killing program.[4] Citing deceptive and misleading official statements regarding the program's success, four former servicemen—Brandon Bryant, Cian Westmoreland, Stephen Lewis, and Michael Haas—claim civilian casualties in US drone strikes inflame anti-American sentiment in the Muslim world, and serve as a recruitment tool, not unlike the detention center at Guantanamo Bay, for extremist groups like ISIS and al-Qaeda. The former airmen conclude: "This administration and its predecessors have built a drone program that is one of the most devastating driving forces for terrorism and destabilization around the world."

Two years earlier, when General Stanley McChrystal told *Reuters*, "The resentment created by American use of unmanned strikes … is much greater than the average American realizes," the one-time commander of US forces in Afghanistan made headlines (Alexander, 2013). But when veterans of the drone wars expressed similar concerns about blowback, commercial and corporate news organizations scarcely mentioned it (Hanrahan, 2015). Such is the double standard political and military elites insist upon, and the Washington press corps dutifully adheres to. Instead, media narratives routinely focus on the novel lifestyle remote control warriors lead (Stahl). The virtual blackout of the servicemen's forceful denouncement of drone warfare in US corporate media further demonstrates the news industry's complicity in legitimizing and normalizing this new kind of war.

In contrast, through ongoing coverage of drone whistleblowers, independent news outlets such as *Common Dreams*, *The Intercept*, and *The Guardian*—provide a platform for the emergence of an alternative order of discourse surrounding the targeted killing program. Likewise, several independently produced documentaries—Robert Greenwald's *Unmanned: America's Drone Wars* (2013), Tonje Schei's *Drone* (2014) and Sonia Kennebeck's *National Bird* (2016)—feature the insights and perspectives of drone whistleblowers otherwise missing from news and official discourse. All told, whistleblower testimony challenges a dominant narrative that mobilizes the sublime appeal of robotic war to rationalize targeted killing and obscure the strategic fallacy underpinning America's drone campaign.

Pegged to the New York City premier of *Drone*, the open letter's authors appeared together for the first time in an exclusive interview on Pacifica's daily newscast *Democracy Now!* The four men repeated and amplified their criticism of the drone program—despite the prospect of government retaliation under the Obama administration's "war on whistleblowers" (Prupis, 2016; Van Buren, 2012). That is to say, like former NSA officer Thomas Drake, US Army PFC Chelsea Manning, and NSA contractor Edward Snowden before them—all of whom have been charged under the Espionage Act—the airmen face potential prosecution for revealing details about the secretive program.[5] Jessica Radack, a former ethics adviser to the US Department of Justice who is representing the airmen, told reporters, "They're taking an enormous and very brave public risk in speaking out. … not because they've done anything wrong or even revealed classified information, which they're not here to do today, but because they have embarrassed the US government" (Goodman & Gonzalez, 2015).

Radack, whose clients include former national security and intelligence officers who likewise went public with allegations of government fraud, mismanagement, and abuse, explained the former servicemen's motivations this way:

> We're hoping that today, by going public, that this will have more of an influence in the debate, because somehow there's a complete disconnect between these terrorist attacks in Paris and elsewhere and the fact that the drone program has fueled ISIS and al-Qaeda and a number of terrorist groups, and that really needs to be addressed. (Goodman & Gonzalez, 2015)

Like the witnesses we met in preceding chapters, drone whistleblowers aim to instigate and sustain robust debate over the legal, strategic, and ethical implications of robotic warfare. Speaking to *The Guardian* following her testimony before the European parliament alongside fellow whistleblower Cian Westmoreland, Lisa Ling highlighted the political significance of bearing witness to robotic war: "As citizens we need to have some conversation about the things that are in the dark … The people who are out of the picture are the people on the ground within the drone program, and the victims" (Ross, 2016). But as we noted in Chapter 3, policy debates over the US drone campaign ebb and flow, rarely gaining traction with either lawmakers or news workers. Even during the high-stakes and highly contested 2016 Presidential campaign, the drone program was a non-issue (Howley, 2016).

When he took office in 2009, Barack Obama assumed the reins of George W. Bush's drone program. During his tenure, Obama supervised the growth of a sprawling bureaucracy that institutionalized targeted killing as a matter of US legal, military, and foreign policy (Jaffer, 2016). And despite last minute calls for the President to create safeguards that would prevent his successor, Donald Trump, from further expanding the drone program, Obama failed to do so (Devereaux, 2016). Within days of taking office, Trump, whose authoritarian tendencies alarm many people in and outside of government, reauthorized the CIA to conduct drone strikes without Pentagon or White House approval (Sinkman, 2017). As for the former drone operators turned whistleblowers, the lack of meaningful public debate over America's targeted killing program adds insult to injury.

The Rise and Fall of Remote-Control Warriors

In early 2016, the Defense Department announced a new designation—Direct Impact on Combat Operations—to acknowledge the wartime efforts of drone pilots and other "cyber technicians" (McCleary, 2016). Three years earlier, then-Defense Secretary Leon Panetta caused an uproar among veterans groups and lawmakers when he created the Distinguished Warfare Medal to recognize the "extraordinary achievement" of this new class of remote-control warriors. Under pressure, the Pentagon backed off plans for issuing medals to drone operators, pending further review. The outcome of that review created the new award, but fell short of endorsing medals for remote crews.

The controversy over issuing commendations for drone operators reflects a deep ambivalence toward robotic warfare within the US armed forces. Fighter pilots jokingly refer to their counterparts in the targeted killing program as the "Chair Force." Likewise, claims that drone operators are little more than glorified video game enthusiasts are common in both the military and popular culture (Singer, 2009; Stahl, 2013). All of which leads some drone operators to consider themselves second-class citizens. In one of his earliest public statements, former sensor operator Brandon Bryant told reporters:

> The United States government has not really done a good job of humanizing the people that do it. And everyone else thinks that the whole program or the people behind it are a joke, that we are video-game warriors, that we're Nintendo warriors. And that's—that's really not the case. (Goodman & Gonzalez, 2013)

The secrecy and sanitized news coverage surrounding the targeted killing program fosters this animosity and contempt. As a result, drone operators grow increasingly alienated from their colleagues, friends, and family. Former technician turned whistleblower, Lisa Ling, described the psychological dimensions of killing by remote control as "a schizophrenic existence between war and peace." Adding, "the distance between what I was doing and the people I knew was a huge chasm" (Ross). Unsurprisingly, as the Obama administration escalated the drone campaign, the attrition rate for drone operators soared. According to an internal Air Force memo obtained by *The Daily Beast*, the drone fleet was "at a breaking point" (Majumdar, 2015). In turn, the Pentagon began offering salary incentives to recruit new personnel and retain experienced drone operators (Schogol, 2015).

Between the stress of working twelve hour shifts for six and seven days a week, the cognitive dissonance associated with commuting to and from the (virtual) battlefield, and the traumatic experience of witnessing drone strikes, it is no wonder these remote-control warriors have been described as "the lost generation" (Chatterjee, 2015). Consider, for example, Cian Westmoreland's vivid and unsettling description of nightmares related to his experience as a communication technician for the targeted killing program.

> I'm in the radio unit flipping switches, with my boss yelling at me to get it up and running. Then all of a sudden it does start working and I realize with a jolt what I've done. I run out of the control station and now I'm in village in Afghanistan and the whole place is burnt out and there's a woman on the ground covered in soot and a child crying over here. I go up to help the child, but half of her face is blown off and there's nothing I can do about it. (Pilkington)

This and other equally disturbing whistleblower accounts reveal the trauma engendered by the distributed intimacy of drone warfare. Put differently, the sublime appeal of killing by remote control comes at a tremendous cost for the service men and women who conduct these lethal assaults. Operations conducted in secret, with ever increasing frequency, and with little public scrutiny or oversight. In this light, whistleblowers render an altogether different service: exposing the hidden costs of robotic war for those who actually do the killing. Fittingly, the technological sublime inspires a "work around" for drone operators' PTSD: the US Air Force plans to introduce man-machine interfaces—akin to virtual assistants—that will fire weapons on voice command; ostensibly relieving remote-control warriors of the guilt and trauma produced by robotic war (Gonzalez, 2013).

Finally, whistleblower testimony offers important insights into the relationship between technology and the psychosocial implications of this new kind of war. In a chilling sequence from the documentary *Drone*, former sensor operator Michael Haas offers a disturbing glimpse into the mindset cultivated by drone warfare: "Ever step on ants and never give it another thought? That's what you are made to think of the targets—as just black blobs on a screen" (Løge & Schei, 2014). Haas's description recalls a passage from Leo Marx's (1964) seminal work on the force of the technological sublime in American culture. Marx relates a passage from Nathaniel Hawthorn's journal, wherein the esteemed writer, temporarily jarred from his pastoral idyll by a passing train, reflects on the psychological menace inspired by this new technology.

> With the train out of earshot and quiet restored, Hawthorne continues his observations. An ant colony catches his eye. …

> Then "like a malevolent genius," he drops a few grains of sand into the entrance of an ant hole and obliterates it. The result is consternation among the inhabitants, their frantic movements displaying their "confusion of mind." How inexplicable, he writes, must be the agency which has effected this mischief. (1964, p. 14)

Hawthorn's rumination on the railroad, what David Nye (1994) describes in terms of "the dynamic sublime," finds a twenty-first century corollary in the use of armed drones in America's decades-long war on terror: a perpetual war shrouded in secrecy, conducted by remote control, and waged now by three US presidents (Sharp, 2017). Herein lies the value and importance of whistleblowers bearing witness to drone warfare. That is to say, advocates routinely employ the rhetoric of the technological sublime to rationalize targeted killing in the name of national security. But as Carey and Quirk observe, this discourse "constitutes a false consciousness" (1992, p. 120) that compromises our ability to fully consider the legal, strategic, and ethical implications of new technologies—let alone the power relations articulated by such material-discursive practices. With their intimate testimony, drone whistleblowers challenge the dominant narrative that justifies, sanitizes, and normalizes targeted killing. Equally important, these and the other witnesses we have met throughout this section offer instead an alternative order of discourse: one that opens up discursive space for dissent and the expression of popular resistance to robotic war. We take up this oppositional order of discourse in the chapters that follow.

Notes

1. The story was subsequently updated in August 2009. See "Drones: America's New Air Force," *CBS News*, http://www.cbsnews.com/news/drones-americas-new-air-force/.
2. Throughout, I use the designation "drone operator" to refer to various personnel involved in conducting drone strikes: remote pilots, sensor operators, and intelligence analysts.
3. I do not use this word lightly. Speaking to *The Guardian* newspaper, former sensor operator Brandon Bryant described one such attack mission in precisely these terms. "We waited for those men to settle down in their beds and then we killed them in their sleep. That was cowardly murder." See Pilkington (2015).
4. In addition to the President, the letter is addressed to Secretary of Defense, Ash Carter, and CIA Director, John Brennan. See https://www.documentcloud.org/documents/2515596-final-drone-letter.html.
5. In 2015, Brandon Bryant joined the ranks of Chelsea Manning and Edward Snowden, previous winners of Germany's biennial whistleblower award, after revealing the role of Ramstein Air Force Base in the targeted killing program. See "Former US drone operator

to get German whistleblower award," *Stars and Stripes*, https://www.stripes.com/news/former-us-drone-operator-to-get-german-whistleblower-award-1.373428.

References

Alexander, D. (2013, January 7). Retired general cautions against overuse of "hated" drones. *Reuters*. Retrieved from http://www.reuters.com/article/us-usa-afghanistan-mcchrystal-idUSBRE90608O20130107

Asaro, P. M. (2013). The labor of surveillance and the bureaucratized killing: New subjectivities of military drone operators. *Social Semiotics, 23*(2), 196–224.

Banet-Weiser, S. (2012). *Authentic™: The politics of ambivalence in a brand culture*. New York, NY: New York University Press.

Benjamin, M. (2013). *Drone warfare: Killing by remote control*. London: Verso.

Brodmerkel, S., & Carah, N. (2016). *Brand machines: Sensory media and calculative culture*. London: Palgrave Macmillan.

Carah, N. (2017). Algorithmic brands: A decade of brand experiments with mobile and social media. *New Media & Society, 19*(3), 384–400.

Carey, J. W., & Quirk, J. J. (1992). The mythos of the electronic revolution. In J. W. Carey *Communication as culture: Essays on media and society* (pp. 113–141). New York, NY: Routledge.

Chatterjee, P. (2015, March 5). Is drone warfare fraying at the edges. *TomDispatch*. Retrieved from http://www.tomdispatch.com/post/175964/tomgram%3A_pratap_chatterjee,_is_drone_warfare_fraying_at_the_edges/

Chouliaraki, L. (2015). Digital witnessing in conflict zones: The politics of remediation. *Information, Communication & Society, 18*(11), 1362–1377.

Cockburn, A. (2015). *Kill chain: The rise of the high-tech assassins*. New York, NY: Henry Holt.

Cohen, K. (2010). *Never alone, except for now: Mediated collectivity in networks* (Unpublished doctoral dissertation). University of Chicago. Retrieved from http://gradworks.umi.com/3432709.pdf

Cole, D. (2014, May 10). We kill people based on metadata. *The New York Review of Books*. Retrieved from http://www.nybooks.com/daily/2014/05/10/we-kill-people-based-metadata

Devereaux, R. (2016, November 16). Donald Trump is about to inherit Obama's unchecked drone program. *The Intercept*. Retrieved from https://theintercept.com/2016/11/16/donald-trump-is-about-to-inherit-obamas-unchecked-drone-program/

Di Salvo, P. (2014). Blowing the whistle on drone strikes. *Italian Coalition for Civil Liberties*. Retrieved from https://medium.com/@Cild2014/blowing-the-whistle-on-drone-strikes-49b463eb615f

Edmondson, R. (2001). Studying civic culture ethnographically and what it tells us about social capital: Communities in the west of Ireland. In P. Dekker & E. M. Uslaner (Eds.), *The politics of everyday life* (pp. 59–72). London: Routledge.

Franz, N. (2017). Targeted killing and pattern-of-life analysis: Weaponised media. *Media, Culture & Society, 39*(1), 111–121.

Gonzalez, R. (2013, October 28). Psychologists propose horrifying solution to PTSD in drone operators. *Gizmodo*. Retrieved from http://io9.gizmodo.com/psychologists-propose-horrifying-solution-to-ptsd-in-dr-1453349900

Goodman, A., & Gonzalez, J. (2013, October 25). A drone warrior's torment: Ex-air force pilot Brandon Bryant on his trauma from remote killing. *Democracy Now!* Retrieved from https://www.democracynow.org/2013/10/25/a_drone_warriors_torment_ex_air

Goodman, A., & Gonzalez, J. (2015, November 20). Exclusive: Air Force whistleblowers risk prosecution to warn drone war kills civilians, fuels terror. *Democracy Now!* Retrieved from https://www.democracynow.org/2015/11/20/exclusive_air_force_whistleblowers_risk_prosecution

Gregory, D. (2011). From a view to a kill: Drones and late modern war. *Theory, Culture & Society*, *28*(7–8), 188–215.

Grothoff, C., & Porup, J. M. (2016, February 16). The NSA's SKYNET program may be killing thousands of innocent people. *Ars Technica UK*. Retrieved from http://arstechnica.co.uk/security/2016/02/the-nsas-skynet-program-may-be-killing-thousands-of-innocent-people/

Hanrahan, J. (2015, December 4). Learning to love the "drone war." *Consortium News*. Retrieved from https://consortiumnews.com/2015/12/04/learning-to-love-the-drone-war/

Haraway, D. (1997). *Modest_witness@second_millenium.femaleman©_meets_ocncomouse™: Feminism and technoscience*. New York, NY: Routledge.

Howley, K. (2016). Loose cannons: Or the silent debate on drones. *US Election Analysis 2016*. Retrieved from http://www.electionanalysis2016.us/us-election-analysis-2016/section-3-policy/loose-cannons-or-the-silent-debate-on-drones/

Irvine, D. (2012, February 15). Doing the military's dangerous, dull and dirty work. *CNN.com*. Retrieved from http://www.cnn.com/2012/02/15/business/singapore-airshow-drones/

Jaffer, J. (2016). *The drone memos: Targeted killing, secrecy and the law*. New York, NY: The New Press.

Linebaugh, H. (2013, December 29). I worked on the US drone program—the public should know what really goes on. *AlterNet*. Retrieved from http://www.alternet.org/world/i-worked-us-drone-program-public-should-know-what-really-goes

Løge, L., (Producer), & Schei, T. H. (Director). (2014). *Drone* [Motion picture]. Norway: Flimmer Film.

Majumdar, D. (2015, January 4). Exclusive: US drone fleet at "breaking point", Air Force says. *The Daily Beast*. Retrieved from http://www.thedailybeast.com/articles/2015/01/04/exclusive-u-s-drone-fleet-at-breaking-point-air-force-says.html

Marx, L. (1964). *The machine in the garden: Technology and the pastoral ideal in America*. New York, NY: Oxford University Press.

Mazzetti, M., & Schmidt, M. S. (2013, June 9). Ex-worker at CIA says he leaked data on surveillance. *New York Times*. Retrieved from http://www.nytimes.com/2013/06/10/us/former-cia-worker-says-he-leaked-surveillance-data.html

McCleary, P. (2016, January 7). Pentagon's drone pilots get a nod, but no medals. *Foreign Policy*. Retrieved from http://foreignpolicy.com/2016/01/07/pentagons-drone-pilots-get-a-nod-but-no-medals/

Naughton, J. (2016, February 21). Death from above, dished out by algorithm. *The Guardian*. Retrieved from https://www.theguardian.com/commentisfree/2016/feb/21/death-from-above-nia-csa-skynet-algorithm-drones-pakistan

Nye, D. E. (1994). *American technological sublime*. Cambridge: MIT Press.

Pilkington, E. (2015, November 19). Life as a drone operator: "Ever step on ants and never give it another thought?" *The Guardian*. Retrieved from https://www.theguardian.com/world/2015/nov/18/life-as-a-drone-pilot-creech-air-force-base-nevada

Power, M. (2013, October 22). Confessions of an American drone operator. *GQ*. Retrieved from http://www.gq.com/story/drone-uav-pilot-assassination

Priest, D. (2013, July 21). NSA growth fueled by need to target terrorists. *Washington Post*. Retrieved from https://www.washingtonpost.com/world/national-security/nsa-growth-fueled-by-need-to-target-terrorists/2013/07/21/24c93cf4-f0b1-11e2-bed3-b9b6fe264871_story.html?utm_term=.51e895aa0920

Prupis, N. (2016, April 21). Thanks to "war on whistleblowers," US ranks 41st on press freedom index. *Common Dreams*. Retrieved from http://www.commondreams.org/news/2016/04/21/thanks-war-whistleblowers-us-ranks-41st-press-freedom-index

Rentschler, C. (2011). From danger to trauma: Affective labor and the journalistic discourse of witnessing. In P. Frosh & A. Pinchevski (Eds.), *Media witnessing: Testimony in the age of mass communication* (pp. 158–181). Basingstoke: Palgrave Macmillan.

Ross, A. (2016, July 1). Former US drone technicians speak out against programme in Brussels. *The Guardian*. Retrieved from https://www.theguardian.com/world/2016/jul/01/us-drone-whistleblowers-brussels-european-parliament

Rothenberg, D. (2014, November 6). Interview with a US Air Force drone pilot: It is, oddly, war at a very intimate level. *Foreign Policy*. Retrieved from http://foreignpolicy.com/2014/11/06/interview-with-a-u-s-air-force-drone-pilot-it-is-oddly-war-at-a-very-intimate-level/

Rowley, C. (2014, September 30). Beheadings v. drone assassinations. *Consortium News*. Retrieved from https://consortiumnews.com/2014/09/30/beheadings-v-drone-assassinations/

Sadowski, H. (2016). *Digital intimacies: Doing digital media differently*. Linköping Studies in Arts and Science. No. 691. Retrieved from https://liu.diva-portal.org/smash/get/diva2:1047582/FULLTEXT01.pdf

Scharrer, E., & Blackburn, G. (2015). Images of injury: Graphic news visuals' effects on attitudes toward the use of unmanned drones. *Mass Communication and Society*, *18*, 799–820.

Schogol, J. (2015, January 15). Air force raises monthly incentive pay for drone pilots. *Air Force Times*. Retrieved from https://www.airforcetimes.com/story/military/2015/01/15/more-pay-for-drone-pilots/21829659/

Serle, J. (2016 July 21). Infographic: The CIA's drone war in Pakistan. *The Bureau of Investigative Journalism*. Retrieved from https://www.thebureauinvestigates.com/2016/07/21/infographic-cias-drone-war-pakistan/

Sharp, J. (2017, February 10). How drones and robotics may shape the future of conflict under President Trump. *Public Radio International*. Retrieved from https://www.pri.org/stories/2017-02-10/how-drones-and-robotics-may-shape-future-conflict-under-president-trump

Singer, P. W. (2009). *Wired for war: The robotics revolution and conflict in the 21st century*. New York, NY: Penguin Press.

Sinkman, P. D. (2017, March 14). Report: Trump gives CIA authority for drone strikes. *US News*. Retrieved from https://www.usnews.com/news/politics/articles/2017-03-14/report-trump-gives-cia-authority-for-drone-strikes

Soderbergh, S. (Producer), & Poitras, L. (Director). (2014). *Citizenfour* [Motion picture]. New York, NY: Praxis Films.

Stahl, R. (2013). What the drone saw: The cultural optics of the unmanned war. *Australian Journal of International Affairs*, *67*(5), 658–674.

Tait, S. (2011). Bearing witness, journalism and moral responsibility. *Media, Culture & Society*, *33*(8), 1220–1235.

United States Air Force. (2015, October 13). Air Force distributed common ground system. *USAF*. Retrieved from http://www.af.mil/AboutUs/FactSheets/Display/tabid/224/Article/104525/air-force-distributed-common-ground-system.aspx

Van Buren, P. (2012, February 9). Obama's unprecedented war on whistleblowers. *Salon.com*. Retrieved from http://www.salon.com/2012/02/09/obamas_unprecedented_war_on_whistleblowers/

Weaver, S. D., & Gahegan, M. (2007). Constructing, visualizing, and analyzing a digital footprint. *Geographical Review*, *97*(3), 324–350.

Williams, B. G. (2013, May 23). Inside the murky world of "signature strikes" and the killing of Americans with drones. *The Huffington Post*. Retrieved from http://www.huffingtonpost.com/brian-glyn-williams/nside-the-murky-world-of-_b_3367780.html

Williams, J. (2015). Distant intimacy: Space, drones and just war. *Ethics and International Affairs*, *29*(1), 93–110.

Winch, A. (2015). Brand intimacy, female friendship and digital surveillance networks. *New Formations*, *84/85*, 228–245.

Wood, D. (2013, May 15). Drone strikes: A candid, chilling conversation with top US drone pilot. *The Huffington Post*. Retrieved from http://www.huffingtonpost.com/2013/05/15/drone-strikes_n_3280023.html

PART IV

RESISTANCE

·10·

DIRECT ACTION AND MEDIA ACTIVISM

We're trying to alert the public that there's a shadow side to this seduction with drones.
—Ron Faust, PeaceWorks Kansas City[1]

Writing in an open letter to President Barack Obama dated April 12, 2016, US Army chaplain Chris Antal resigned from the military. In words and deeds, the Unitarian Universalist Church minister scorned the doctrine of perpetual war pursued by his commander-in-chief, the Nobel Peace Prize laureate. Citing the Obama administration's targeted killing program, its plans to modernize America's nuclear arsenal, and its policy of global military supremacy, Antal's resignation read in part:

> The executive branch continues to claim the right to kill anyone, anywhere on Earth, at any time, for secret reasons, based on secret evidence, in a secret process undertaken by unidentified individuals. … I refuse to support this policy of unaccountable killing. (Marcetic, 2016)

Like the drone operators turned whistleblowers we met in Chapter 9, Antal told *In These Times*, a non-profit, social justice magazine, of the "moral injury" he suffered from bearing witness to the US drone campaign in Afghanistan. Refusing to serve as an "empire chaplain," Antal expressed his dismay over the conduct of America's wars since September 11, 2001, and questioned blind obedience to what he called "the holy trinity" of the state, the military, and capitalism (Marcetic). Antal's letter received considerable attention in the independent and activist press (Norton, 2016; Whitton, 2016). But like the whistleblowers before him, his forceful condemnation of drone warfare received far less traction in commercial and corporate news media. Mainstream media silence surrounding Antal's opposition to what he once described as "the colossal misery our wars inflict on the innocent"[2] should come as no surprise. As we've noted throughout, American news workers routinely consult, and frequently defer to, official sources—political and military elites—who extol the sublime virtues of killing by remote control. All the while marginalizing the voices of survivors and witnesses of US drone strikes, and neglecting the perspectives of peace activists and veterans groups opposed to America's targeted killing program.

When high profile figures, such as retired General Stanley McChrystal or Pakistani women's rights activist Malala Yousafzai, claim drone attacks fuel anti-American sentiment and drive local communities to violent extremism, they bring critical, albeit short-lived media attention to the relationship between the US drone campaign and violent blowback against America and its allies (Alexander, 2013; Clark, 2013). Likewise, when Rand Paul took the floor of the US Senate to criticize the targeting of American citizens for extrajudicial killing, the ensuing news coverage generated unprecedented public awareness of the classified program. Despite such criticism, however, a sanitized narrative touting the precision, accuracy, and efficiency of US drone strikes remains the dominant order of discourse. And, notwithstanding the alternative order of discourse produced by the victims, survivors, and witnesses we met in the previous section, opposition to the escalation and expansion of America's drone wars is difficult to sustain in mainstream news discourse. It is here, first and foremost, that we can detect the discursive and political significance of growing resistance to drone warfare articulated through nonviolent protest, activist media, and popular culture.

This chapter, the first of three to examine an oppositional discourse to targeted killing, explores direct action campaigns and other grassroots efforts aimed at grounding the drones. Throughout, I highlight the importance of alternative and activist media in the formation of this oppositional order of discourse. Doing so, I reveal the synergies between the alternative order of discourse articulated by witnesses to atrocity, and organized resistance to drone warfare at home and abroad. Indeed, drone whistleblowers have become fixtures at protests, educational workshops, and legislative hearings organized by anti-war activists (Kupfer, 2016; Ross, 2016). Furthermore, I demonstrate the significance of intertextuality and interdiscursive practice in shaping and informing this oppositional discourse. As we shall see, this oppositional discourse draws from news and official statements regarding targeted killing, survivor and witness testimony, as well as the discourses of civil liberties, criminal justice, and racial inequality.

The chapter proceeds as follows. First, I consider the role of organized protest in the formation of this oppositional discourse, with an eye toward evaluating news accounts of anti-drone activism in local, national, and international news media. Following this, I discuss an exemplar of activist media practice—the grassroots educational organization KnowDrones—and the strategies and tactics used by this group to organize popular resistance to America's drone wars. The chapter concludes with a brief discussion of the significance of street theater and "ironic media activism" (Day, 2011) in generating public awareness of the (purposefully) hidden costs and consequences of drone warfare.

Ground the Drones

As noted in Chapter 3, the year 2013 was a watershed in the public life of drones. Following the leak of a classified "white paper" several Congressional oversight committees began pressuring the Obama administration to release the Department of Justice's legal memorandum on the targeting of American citizens. The short-lived breakdown in elite consensus over the classified drone program stalled John Brennan's confirmation as Obama's new CIA director. That same year, victims and survivors appeared before lawmakers for the first time and offered dramatic testimony detailing the civilian toll of America's drone wars. Meanwhile, a handful of drone operators openly challenged official claims regarding the accuracy and precision of US drone strikes, and spoke candidly about the pain and suffering caused by their participation in this new kind of war. All told, increased public scrutiny of the secretive program galvanized a nascent resistance movement to drone warfare.

Opponents of weaponized drones had been active for some time. In 2009, the Upstate Coalition to Ground the Drones and End the Wars (hereafter the Upstate Coalition) began organizing protests outside Hancock Field Air National Guard Base, home of the 174th Attack Wing. Located adjacent to Syracuse International Airport in upstate New York, Hancock personnel pilot MQ-9 Reaper planes in the skies above Afghanistan. That same year demonstrations commenced at Creech AFB in Nevada, the first facility to exclusively fly remotely piloted aircraft. Every year since, a coalition of activists, known as Creechers, organize demonstrations outside that facility (Bernstein, 2016). Finally, the aforementioned activist group, KnowDrones, launched the first of its public awareness campaigns in 2012. But in 2013, a series of critical discourse moments surrounding the escalation and expansion of the drone program under Barack Obama energized anti-drone activists as never before. Put differently, a breakdown in elite consensus, coupled with the testimony of survivors and witnesses, opened up discursive space for anti-drone activism. Seizing the opportunity, activists stepped up their efforts to promote and sustain an otherwise deficient public debate over the legal, moral, and ethical implications of targeted killing. As we shall see, news coverage of these direct action campaigns followed a predictable pattern: one that further demonstrates US corporate media's complicity in legitimating and normalizing drone warfare.

The relationship between protests movements and news organizations is highly fraught and mutually dependent (Boykoff, 2006; Gamson & Wolfsfeld, 1993). For activists, mass media are essential for raising public awareness and communicating their aims and objectives to wider publics. In particular, news media offer activists a valuable resource for organizing resistance and mobilizing constituencies to take action (Ryan, 1991). At the same time, activists are keenly aware of reporters' tendency to trivialize, and sometimes demonize, such protests (Solomon, 2000). Conversely, news workers attracted by the spectacle of public demonstrations, if not the possibility for violent conflict between protestors and the authorities, relish the opportunity to cover such events. Provided of course these protests do not threaten the vested interests of news organizations, their commercial clients, or the governmental agencies and officials that regulate the media industries, and upon whom they rely as information sources. In the discursive struggle over the framing of organized protest, news organizations wield considerable power to shape and sway public opinion (Gitlin, 1980).

Like any number of resistance movements, from anti-war protests of the Vietnam-era to the Occupy Wall Street demonstrations of 2011, anti-drone activists have a complex and complicated relationship with news media. Writing for the Nieman Watchdog Project, John Hanrahan (2011) identifies revealing shortcomings in US news coverage of anti-war protests since Barack Obama took office. Citing public demonstrations outside the White House, the headquarters of major defense contractors, and military installations across the United States, Hanrahan acknowledges the scale of these protests may not match those of previous anti-war demonstrations, but, "the fact is, *protest is much more widespread than citizens might gauge from coverage in newspapers and television* [emphasis added], which seldom report anti-war actions regardless of how significant or newsworthy they may be."

Viewed in this light, media silence surrounding anti-drone activism performs the important ideological work of ignoring or trivializing organized resistance to this new kind of war. In other words, marginalizing dissent in this fashion is part and parcel of "the perception management techniques" (Solomon, 2005) designed to make war easy: a strategy that works to enlist popular support for military intervention, suppress dissent, and otherwise legitimate America's imperial ambitions. Equally important, Hanrahan criticizes US corporate media's refusal to cover the trials of anti-war protestors, thereby denying the public a better understanding of the legal infrastructure put in place in the wake of 9/11: an administrative and judicial regime with little tolerance for dissent of any kind. Hanrahan concludes, "The pattern of the national media is to ignore courtroom trials also, regardless of their drama, leaving them to local news reports or websites." As I aim to demonstrate, the bulk of reporting on anti-drone protests appears in either local outlets or the alternative and activist press. Here then, I want to examine several news accounts of nonviolent direct action targeting the US drone program.

Headlined "Protesters arrested outside upstate NY drone base," an AP wire report published in the Albany *Times Union* describes an anti-drone demonstration at Hancock Field. The story claims 30 of approximately 250 demonstrators were "charged with minor violations" for blocking traffic following the expiration of a protest permit. Although Sheriff Kevin Walsh is mentioned by name, neither he nor any of the protestors are quoted directly. The report duly notes that the Upstate Coalition organized the protests and briefly summarizes the group's concerns. Nevertheless, the story's awkward conclusion inadvertently underscores the expansion of America's drone campaign above declared and undeclared battlefields: "Drones have increasingly been used in Iraq and Afghanistan and strikes in Pakistan" (Associated Press, 2011). Notwithstanding the story's brevity, the framing clearly centers on the "illegality" of the protest, rather than the motivations behind it. What's more, the reference to "minor violations" suggests the arrests themselves were rather trivial. The expiration of a protest permit indicates that demonstrators had, for a time, complied with increasingly draconian restrictions on freedom of speech and assembly—a significant but frequently overlooked issue in news coverage of organized protests. And yet, this *deviance frame* dominates the narrative. A cursory examination of headlines from regional news outlets illustrates the preeminence of this framing.

> **Officers take 16 into custody this morning at Hancock Field drone protest**
> (*Syracuse.com* October 25, 2012)
>
> **Albany man arrested at drone protest**
> (*Times Union* April 29, 2013)
>
> **Jail time, fines in drone protest**
> (*Times Union* February 12, 2014)
>
> **Another drone protestor is convicted in DeWitt court**
> (*Syracuse.com* July 18, 2014)

Taking this deviance frame to extremes, one report posted to Syracuse.com, the online edition of the Syracuse *Post-Standard*, lists the name, age, and home address of 30 people arrested at the April 29, 2013 demonstration (Mulder, 2013). Even by contemporary journalistic standards, such tactics seem excessive: a thinly veiled effort, perhaps, to shame protestors in the court of public opinion. Conversely, the report rather uncharacteristically avoids direct quotes from local law enforcement. Rather, several protestors have their say—albeit briefly. Significantly, activists not only address the immorality of drone warfare, but also question legal restrictions placed on anti-drone activists (Breidenbach, 2013). For Elliott Adams, a protest organizer, the local judiciary's recent decision to grant an "order of protection"—a restraining order commonly used in domestic violence cases—for base commander, Colonel Earl Evans, abridges demonstrators' freedom of speech and assembly. Calling the ruling "absurd," Adams concludes, "It is a flagrant effort to deny us our first amendment rights" (Mulder).

A related storyline follows the arrest, conviction, and imprisonment of Mary Anne Grady Flores, a member of the Upstate Coalition, jailed for photographing a demonstration outside Hancock. Charged with violating the order of protection prohibiting demonstrators from approaching the home or workplace of Col. Evans, Grady Flores vigorously defended her first amendment rights in a series of appellate hearings. Explaining her actions on the eve of her incarceration, the grandmother of three praised drone whistleblowers for exposing the criminality of targeted killing and inspiring civilians to speak out against drone warfare (Goodman, 2016). The prospect of the "anti-drone granny" doing time simply for photographing the Upstate Coalition's actions made her something of a cause célèbre (Moyers, 2016). Equally important, as Brian Terrell of Creative Voices for Nonviolence observed, the case brought renewed attention to the discursive gymnastics underpinning the Obama administration's legal rationale for its targeted killing program—and the implications these language games have for organized resistance. Terrell told *In These Times*:

> The drone program is so blatantly illegal that the authorities have to go to absurd lengths to justify and protect it. They need to redefine words like "imminent threat" and "due process" to cover up the criminality. Here, they have to redefine the meaning of an Order of Protection. I hope that this will bring more people out to protest. (Berrigan, 2014)

While the deviance frame continues to inform a great deal of local news coverage, over time this coverage has grown somewhat sympathetic toward anti-war activists. Consider this curious item from the Springfield *News-Leader*. Headlined "Activists to protest drone attacks," the report previews an upcoming demonstration outside Whiteman Air Force Base in Knob Noster, Missouri. The news story introduces readers to Vicke Kepling, director of the Peace Network of the Ozarks, and offers this description of the protestors' objectives:

> The demonstration, organized by activists in Kansas City and Springfield, will aim to highlight the legal and moral questions around this kind of warfare in which pilots stationed at Whiteman and other US military bases use remote-controlled drones to drop Hellfire missiles or other weapons on alleged terrorists in Afghanistan, Pakistan and other countries. (Shesgreen, 2016)

The report includes brief statements by Kepling and Roger Faust, another anti-drone activist, describing their motivations for organizing the demonstration. Despite the reporter's willingness to give activists a public forum to register their opposition to drone warfare, the balance of the story offers unambiguous support for targeted killing. *USA Today* contributor Deirdre Shesgreen writes:

> There's no question that drones, also known as unmanned or remotely piloted aircraft, have transformed modern warfare. Supporters, including Springfield-area lawmakers, say these weapons provide a super-accurate way to kill America's enemies while minimizing risks to U.S. military personnel. (Shesgreen)

Here, the reporter echoes official sources while invoking the sublime appeal of robotic warfare: its precision in taking out the bad guys while keeping America's fighting force out of harm's way. Following this, Shesgreen gives US Senator Claire McCaskill (D-MO) further opportunity to defend the practice of targeted killing. McCaskill claims, "We have taken out the top levels of leadership in ISIS with a very few civilian casualties." Repeating President Obama's assurances of the surgical precision of US drone strikes, McCaskill adds, "Having been to Whiteman and watched very closely what they do, I think people don't realize what goes into determining a target and what pinpoint accuracy is involved." McCaskill concludes that anti-drone activists, "Just don't have the facts" (Shesgreen). Well intentioned though they may be, the story implies, opponents of drone warfare are in no position to judge the secretive program's efficacy, let alone its legal, ethical, and moral consequences.

Mindful of her constituency, McCaskill acknowledges the burden this new kind of war places on remote pilots. Here, the story reiterates concerns the Senator expressed in a letter to the Pentagon regarding the health and well being of drone operators at Whiteman.[3] McCaskill writes: "They could be playing with their children shortly after witnessing up close and in graphic detail the effects of a 500-pound bomb or Hellfire missile on a soft target." Here, the euphemism "soft target" belies McCaskill's indifference to the fate of people on the receiving end of US drone strikes, while signaling her profound concern for the health and well being of American military personnel. As a counterpoint, Shesgreen adds a note of reassurance, citing the Air Force's recent decision to reclassify drone operators as attack squadrons: "a designation with greater prestige that could also give those personnel access to more resources." McCaskill's support for the troops is, of course, laudable. But as we noted in Chapter 9, this sort of administrative fix fails to substantively address the emotional and psychological trauma drone operators must endure; let alone the specious legal justifications behind the targeted killing program. Nonetheless, the news story uncritically endorses McCaskill's position. The report's accompanying photograph, a headshot of the senator, clinches the newspaper's none-too-subtle effort to reorient the story away from the legal and moral objections to drone warfare raised by peace activists. In short, a news report ostensibly about local opposition to drone warfare sidelines the protest and instead bolsters the Obama administration's rationale for escalating and expanding the US drone campaign.

In contrast, independent and alternative news outlets offer a far more comprehensive and nuanced understanding of nonviolent struggle against drone warfare. Doing so, these reports provide valuable insights into organizing practices that build and sustain resistance movements.

For example, a report in the socialist news outlet, *Workers World*, pegged to an April 27, 2014 demonstration outside Hancock Field, notes, "the multinational march was part of a regional day of education and action linking poverty, racism and war" (Pratt, 2014). To that end, demonstrators carried images of Pakistani victims of US drone strikes, while others held posters of Trayvon Martin and Troy Davis, two African American men executed by a Florida vigilante and the state of Georgia, respectively. Through visual rhetoric, activists link two seemingly disparate forms of state-sponsored terrorism: drone strikes across the Muslim world and the systemic racism of American law enforcement (Kelly, 2016). Equally important, taking a page from the government's playbook, protestors re-appropriate judicial language used to suppress dissent by issuing a "People's Order of Protection Against Drone Terrorism." Activists delivered the order to the 174th Attack Wing in a symbolic yet altogether meaningful effort to end indiscriminate bombing of civilians in America's drone wars.

Unlike related stories in mainstream media focusing on the spectacle of public demonstrations, the *Workers World* report highlights the educational component of the day's action. Recounting an address delivered by Dr. Cornell West, the renowned theologian and social justice activist, the story references portions of West's speech illuminating the relationship between imperialism, racism, and economic inequality. West told the standing-room only crowd at Tucker Missionary Baptist Church: "There is always money for wars and for prisons but not for health care and education" (Pratt). Eschewing the deviance frame that characterizes local news coverage of anti-drone protests, this report offers instead a version of events which underscores the discursive strategy of organizing resistance to inter-linked systems of oppression. Even when the alternative press employs a deviance frame—as in an item from *The Progressive* magazine headlined "Peace activists arrested in anti-drone protests"—the report invariably focuses on the criminality of US drone strikes, the coalition politics of anti-drone activism, and the strategies and tactics activists employ to resist endless war (Kupfer). In contrast to corporate media accounts designed to delegitimize protest, independent and activist media address readers as agents capable of resisting oppressive regimes of power.

Like the drone program itself, local opposition to this new kind of war has assumed global proportions. In recent years, Royal Air Force (RAF) base Waddington in Lincolnshire, England, has been the site of a series of protests calling for an end to drone warfare. Unsurprisingly, US corporate news organizations rarely pick up these stories. Thus, while news outlets and activist sites across England ran the story, the progressive news site, *Common Dreams*, was one of the few American publications to cover the January 5, 2015 arrest of four anti-drone activists who used wire cutters to make their way into RAF Waddington through what they called a "New Year gateway to peace." Once inside, the four unfurled a banner reading, "End the Drone Wars," and brandished published reports documenting civilian casualties of coalition airstrikes in Afghanistan and Iraq. Base security quickly apprehended the activists, charging them with aggravated trespass. Nonetheless, the small band of protestors made their point.

Drawn from reporting in *The Guardian* and the BBC, *Common Dreams* published excerpts of a joint statement issued by the demonstrators. It read in part:

> Thanks to the marketing of drone war as "risk free," "precise" and above all "humanitarian," war has been rehabilitated and accepted as virtually normal by those who see little or nothing of the impact on the ground thousands of miles away. … But behind the rebranding, war is as brutal and deadly as it has always been with civilians killed, communities destroyed and the next generation traumatized. (Queally, 2015)

In forceful and eloquent fashion, this statement upsets the rhetoric of the technological sublime, common to news and official discourse, lauding the accuracy and precision of drone strikes. Furthermore, the statement's close attention to the language of marketing and branding suggests a keen understanding of the discursive strategies used to "sell" drone warfare to the public. Finally, by reasserting the brutal realities of armed conflict, especially its toll on civilians, the statement reveals what American anti-war activist Ron Faust refers to as "the shadow side of this seduction with drones" (Shesgreen). Notwithstanding the domestic news blackout, mounting protests at RAF Waddington further underscore the discursive and political value of grassroots efforts aimed at grounding the drones. Nowhere have the international dimensions of embodied resistance to drone warfare been made more evident than an escalating series of demonstrations outside Ramstein Air Force base in Germany.

Significantly, anti-drone activism across Germany increased dramatically following former drone operator Brandon Bryant's revelation that Ramstein serves as a coordination hub for America's targeted killing program. Ahead of his testimony before a special committee of the German parliament—yet another story that went largely unreported in US corporate media—Bryant told journalists: "Ramstein is absolutely essential to the US drone program.

All information and data go through Ramstein. Everything. For the whole world. Also for the CIA operations" (Barreto & Kepmann, 2015). Once again, whistleblower disclosures of the inner workings of the American drone campaign prompted direct action (Press TV, 2015). The following year, thousands rallied outside the controversial air base in one of the more remarkable anti-drone demonstrations to date. According to Deutsche Welle, the international public broadcaster, "around 5000 protestors created a human chain close to Ramstein, the principle US Air Force facility in Europe" (Deutsche Welle, 2016).

Despite rising public opposition to the German government's complicity in America's global killing program, this story also failed to gain much traction in US commercial and corporate media. In light of the media blackout surrounding this and a growing number of anti-drone protests (Project Censored, 2016), peace activists, veterans groups and their allies have taken it upon themselves to make their own media, with the aim of grounding the drones through grassroots organizing, public demonstrations, and strategic communication campaigns. Few groups are more effective at this DIY approach to publicizing anti-war activism than KnowDrones.

Activist Media Practice

The KnowDrones tour opened on April 12, 2012. Taking to the streets of Brooklyn, New York, with a replica MQ-9 Reaper, organizers launched their campaign ahead of a public forum later that week featuring Medea Benjamin, co-founder of Code Pink, Glen Ford of the Black Agenda Report, and Zohra Ahmed of the Pakistani Solidarity Network, among others. According to a KnowDrones press release, the tour began in Brooklyn because it is the home district of Representative Edolphus Townes (D-NY), a member of the Congressional Unmanned Systems Caucus—better known as the "drone caucus" (KnowDrones, 2012). In the following weeks, KnowDrones organizers and local activists visited 17 congressional districts across the United States, with stops in Baltimore, Philadelphia, Ohio, and California. The organization's founder, Nick Mottern, a Navy veteran, said: "The goals of this tour are to end US drone attacks and US wars; and to achieve an international ban on weaponized drones and drone surveillance" (KnowDrones). In addition to listing the tour's many co-sponsors, including the War Resisters League, Occupy Wall Street, and Voices for Creative Nonviolence, the press release includes a link to a video featuring raw footage of US drone strikes, as well person-on-the-street interviews and public reaction to the replica aircraft as it made the rounds outside the Museum of Modern Art.[4] Mottern observed, "The drone replicas are powerful teaching tools."

The educational focus of KnowDrones underscores the decisive role government secrecy, selective leaks, and America news workers' official source stenography play in securing popular consent to the drone wars. Like its coalition partners in the peace movement, KnowDrones aims to challenge news and official discourse celebrating the surgical precision of US drone strikes, and instead promote greater understanding of the legal, ethical, and moral consequences of drone warfare. The following year, KnowDrones organized a month-long campaign dubbed "April Days of Action 2013." Designed to "generate a public uprising across the United States to stop drone spying and drone warfare" (Anon, 2013) this latest effort capitalized on the organization's previous success while expanding the group's reach and influence in the anti-war movement. In the weeks leading up to the days of action, KnowDrones urged local organizers from across the country to educate their communities and mobilize popular resistance to drone warfare. In addition to coordinating this nation-wide campaign, the group's website offered community organizers educational resources, including primers on drone technology, legal and policy analyses of targeted killing, independent reports on civilian casualties, and documentation of domestic law enforcement's growing reliance on UAVs.

In this respect, the KnowDrones website operates as an information clearinghouse for groups and individuals to educate themselves on the threats to innocent civilians, national security, and civil liberties posed by these unmanned surveillance and weapons systems. Accordingly, KnowDrones identified three main targets for its month-long action: the manufacturing plants of defense contractors; universities that conduct research and development on unmanned systems; and military facilities that conduct drone US strikes. Speaking to *The Guardian* newspaper on the eve of the campaign, Mottern said: "Most people here in the United States know a little bit about drones; we want everyone to begin to see the depth of the threat that drones present to all of us, regardless of what nation we live in" (Harris, 2013). Predictably, the April Days of Action received limited media attention, primarily in local news outlets. Nevertheless, through public demonstrations and teach-ins, the campaign promoted greater public awareness and participation in anti-drone activism. Likeminded efforts, most notably the November 2013 drone summit organized by Code Pink, featured a rally in front of the White

House and a die-in outside the DC headquarters of General Atomics, manufacturer of Predator and Reaper aircraft (Lazare, 2013).

KnowDrones continues to promote organized resistance to drone warfare through education and outreach programs. For instance, in 2015 the group launched a media campaign designed to persuade drone operators to end their participation in the targeted killing program. Once again, national media paid little mind to anti-drone activists, leaving it to local, independent, and as it happens, military news outlets to cover this latest effort. Partnering with local veterans groups across the country, KnowDrones purchased airtime for a series of 15-second advertisements in television markets nearby US air bases. The sometimes-graphic spots implore drone operators: "Please refuse to fly. No one has to obey an immoral law" (Schogol, 2015). Knowing full well the risks military personnel face for refusing orders, veterans groups nonetheless hope the ads will persuade drone operators to reconsider their role in targeted killing. Thus, the television spots target two distinct audiences: the general public, whose knowledge of the drone program is limited to a sanitized narrative common to news and official discourse; and drone operators, who have intimate knowledge of the brutal reality of drone warfare.

Writing in *The Guardian*, Amanda Holpuch (2015) notes: "One of the ads features images of wounded and dead children, and will appear on television only after 10 PM. The other, edited version will run from 5 AM–9 PM." In early 2015, KnowDrones began running the ads on local television stations in Las Vegas, home to Creech AFB. Soon thereafter, the group purchased airtime in Sacramento, California, nearby Beale AFB. And in May, the Upstate Coalition, in partnership with KnowDrones and the Syracuse Peace Council, began running spots on Fox News, CNN, MSNBC, as well as the YES television network during New York Yankees baseball games (Weiner, 2015). Placed strategically in news and sports programming, the ads aim for maximum exposure to its target audience. In 2016, KnowDrones reprised the ads in Las Vegas to coincide with the trial of 14 activists arrested for demonstrating outside Creech AFB. But in its most audacious move, a year earlier, KnowDrones ran a print ad in the *Air Force Times*, a weekly publication serving current and former Air Force and National Guard members.[5]

Like the television spots, Iraq Veterans Against the War, Veterans For Peace, and World Can't Wait endorsed the print advertisement. Complete with the name, rank, and service affiliation of 54 signatories, the remarkable text begins: "As retired and former members of the US military, we urge US drone pilots, sensor operators and support teams to refuse to play any role in drone surveillance/assassination missions." Arguing that the drone program violates international law and denies individuals due process, the ad relates statistics compiled by the Bureau of Investigative Reporting on civilian casualties across the Muslim world. Invoking the Principles of International Law, the ad concludes that service members have the right—and the responsibility—to refuse unlawful commands. In boldface type, the tag line reads: "So, yes, you do have a choice—and liability under the law. Choose the moral one. Choose the legal one." Drawing on the gravitas of retired military, the advertisement's persuasive appeal operates on several levels: first, to the proud tradition of America's fighting forces; second, to the legal responsibilities of military personnel; and finally, to each soldier's personal morality. A bold and provocative intervention, the ad offers a respectful yet compelling challenge for drone operators to contemplate the legality of targeted killing and the moral hazard thereof.

Creative Resistance

While some activists purchase advertising and airtime to communicate with wider publics, others employ stunts and elements of political theater to "beg, borrow or steal the media spotlight" (Day, p. 145). For example, Code Pink, the women-led activist organization, frequently stages mock drone attacks on weddings and funerals. These elaborate stunts are designed to attract media coverage and incite public indignation over the loss of innocent lives in America's drone wars.

In October 2010, the San Francisco chapter of Code Pink organized a counter-spectacle to a Bay-area performance by the Blue Angels, the US Navy's celebrated aerial demonstration team. Aimed at undermining this celebration of American technological prowess and military might, the action called on local residents to participate in a mock drone attack. The anti-war group's "invitation" read in part: "With a history of recent attacks on wedding parties and funerals … drone warfare symbolizes the total disregard for human life that US war making represents. Can you join us to dramatize this tragedy?" (Code Pink, 2010).

In similar fashion, Code Pink's Washington DC chapter staged a mock drone attack on a wedding party outside the White House. Pegged to the December 2013 drone strike in Yemen discussed in Chapter 6, a video recording of the group's theatrics went viral (NBC, 2014). Here, then, organizers draw on news and official discourse that grudgingly acknowledges civilian casualties when "errant" airstrikes slaughter dozens of innocent people. While such episodes cause military and political leaders some embarrassment, seldom do these same officials assume responsibility for the carnage. Loathe to concede even more despicable crimes, official sources studiously avoid any mention of the controversial tactic known as the "double tap," wherein drones linger over the site of an airstrike only to fire again when first responders arrive at the scene. Code Pink's street theater underscores the cruelty of this technique.

Dramatizing "collateral damage" in this fashion, activists seek first to enrage audiences with the rampant criminality of drone warfare, and then to engage the public in a more forthright debate over the targeted killing program and its consequences. Significantly, Code Pink's discursive strategy of familiarizing the terror of drone warfare rests on re-contextualizing such brutality for American audiences. To that end, organizers encourage participants to wear white "or other wedding-like attire" and to bring a change of black clothes for the funeral. "We'll have drone replicas 'flying' overhead, small coffins, and ragdoll babies to dramatize the child victims" (Code Pink). During the mock attack, participants pretend to be victims of US drone strikes in an effort to make the appalling reality of these assaults palpable for an American public

Figure 10.1: Peace activists stage a mock drone assault on a funeral procession outside Creech Air Force Base in Nevada. *Photo credit: Mauro Oliveira. Source: Code Pink.*

otherwise lulled into complacency by a sanitized narrative that celebrates the cleanliness and efficiency of "surgical" strikes (Figure 10.1).

In contrast to the solemn tone Code Pink's media activism takes, two final examples illustrate the political relevance of activist practice in a more humorous register. Consider, for example, the work of the California Department of Corrections (CDC). Dedicated to "correcting" all manner of public communication, the group's name reflects its ironic take on political art and activism. Like legions of culture jammers before them, the CDC appropriates, disrupts, and upends advertising messages for political ends. On Election Day 2013, the group hijacked a series of advertisements for smartphones adorning municipal bus shelters around San Francisco. Offering sly commentary on the US drone program, the "corrected" ads show a cellphone with the image of a Predator drone launching a Hellfire missile. Above, the word "Pakistan" appears where the company logo should be. The tag line reads: "The Next Big War is Already Here."

For unsuspecting commuters, the corrected smartphone ad provides greater insight into the government's clandestine killing program than a month's worth of Pentagon briefings. Should the CDC's subtlety be lost on puzzled bus riders, the group's website explains its prank in appropriately satirical terms: "As these [drone] operations are shrouded in secrecy, the California Department of Corrections released the rehabilitated smartphone ads to assist our colleagues in the federal government and explain the benefits of drones to war-weary Americans" (Swan, 2013). Assuming the voice and public service mission of an authentic government agency, the CDC draws upon the authority of the state to offer a sly corrective to the lack of government transparency and accountability in its targeted killing program.

Then there is the Drone Survival Guide, created by Dutch artist Ruben Pater, which offers another irreverent, but altogether relevant public service in the age of drone warfare. Emblazoned with ominous images of all manner of militarized drones, complete with country of origin and rendered in silhouette against an aluminum background, the survival guide, currently available in 17 languages, offers readers practical tips for identifying, eluding, and even hacking into the guidance systems of these "robotic birds." According to the guide: "Our ancestors could spot natural predators from afar by their silhouettes. Are we equally aware of the predators in the present-day?" (Pater, 2013). At first blush, likening drone spotting to twenty-first century bird watching may seem fanciful. Nonetheless, the Drone Survival Guide, like the direct action campaigns and media activism discussed above, helps raise public awareness of the imminent threat posed by this new kind of war.

***

In a pointed and dramatic reversal, Pakistani artists have turned American drone operators' unblinking stare into a forceful recrimination of robotic warfare (Schonfeld, 2014). The campaign, dubbed #NotABugSplat, is a reference to military slang for victims of US drone strikes. Voicing their opposition to indiscriminate bombing with equal measures creativity and ingenuity, a local artists collective in Pakistan's remote Khyber Pakhtunkhwa Region, working in collaboration with international partners, produced a giant poster of a young girl, said to have lost her parents and siblings in a US drone strike. According to organizers of the participatory arts project, Inside Out, the installation aims to "show the drone operators the actual face of the children they might end up bombing—rather than a dehumanized dot on a screen" (Inside Out, 2014). Like KnowDrone's television ads, albeit in a slyly subversive fashion, the installation addresses drone operators directly while raising public awareness of the slaughter of innocence in America's drone wars. According to a video produced by the UK-based human rights organization, Reprieve, the hashtag #NotABugSplat went viral, creating an international sensation "with earned media estimated at over $2 million" (Reprieve Foundation, 2014). Despite the campaign's enviable reach, however, its long-term influence appears negligible. All the same, opposition to America's targeted killing program persists in all manner of cultural expression. In the next chapter, we take a closer look at the intersection of graphic arts, digital culture, and popular resistance to drone warfare.

Notes

1. Faust is one of several activists quoted in a local news report about an anti-drone protest outside Whiteman Air Force Base in Missouri. See Shesgreen (2016).
2. Antal made the remarks in his 2012 Veteran's Day sermon. The chaplain's address led to disciplinary action, including an official reprimand and release from active duty (Marcetic). See "A Veteran's Day confession for America," *Quest for Meaning*, https://www.questformeaning.org/military-2/a-veterans-day-confession-for-america/.
3. Helpfully, the online version of this story includes a link to this earlier report (Shesgreen, 2015).
4. The museum was chosen as the site of the sidewalk display to coincide with an exhibition featuring the work of avant-garde filmmaker Harun Farocki titled "Images of War (at a distance)." See "Drones: Less Distance from War." KnowDrones, https://www.youtube.com/watch?v=SbM2mZ8uYBw.
5. Several outlets duplicated the ad. For example, see "Air Force Times runs veteran's ad urging drone pilots to refuse to fly," *Truthdig*, http://www.truthdig.com/eartotheground/item/air_force_times_runs_veterans_ad_urging_drone_operators_20150919.

References

Alexander, D. (2013, January 7). Retired general cautions against overuse of "hated" drones. *Reuters*. Retrieved from http://www.reuters.com/article/us-usa-afghanistan-mcchrystal-idUSBRE90608O20130107

Anon. (2013). Stop US drone warfare. *World Can't Wait*. Retrieved from http://www.worldcantwait.net/images/pdfs/%20Stop_Drones_2up_A.pdf

Associated Press. (2011, April 22). Protesters arrested outside upstate NY drone base. *Times Union*. Retrieved from http://www.timesunion.com/news/article/Protesters-arrested-outside-upstate-NY-drone-base-1349095.php

Barreto, D., & Kepmann, A. (2015, October 15). Brandon Bryant: Ramstein is absolutely essential. *Das Erste*. Retrieved from http://daserste.ndr.de/panorama/aktuell/Brandon-Bryant-Ramstein-is-absolutely-essential,drohnen252.html

Bernstein, D. (2016, June 8). Calling out drone war as a crime. *Consortium News*. Retrieved from https://consortiumnews.com/2016/06/08/calling-out-drone-war-as-a-war-crime/

Berrigan, F. (2014, August 13). Judge jails anti-drone granny. *In These Times*. Retrieved from http://inthesetimes.com/article/17042/judge_jails_anti_drone_granny

Boykoff, J. (2006). Framing dissent: Mass media coverage of the global justice movement. *New Political Science*, 28(2), 201–228.

Breidenbach, M. (2013, April 17). 174th Attack Wing has order of protection from anti-drone peace activists. *Syracuse.com*. Retrieved from http://www.syracuse.com/news/index.ssf/2013/04/174th_attack_wing.html

Clark, L. (2013, October 11). Malala Yousafzai tells Obama drones are "fueling terrorism." *McClatchy*. Retrieved from http://www.mcclatchydc.com/news/politics-government/article24757099.html

Code Pink. (2010). Real angels don't kill people: Mock wedding and funeral drone attack GG bridge noon. *Indybay.org*. Retrieved from https://www.indybay.org/newsitems/2010/10/09/18660997.php

Day, A. (2011). *Satire & dissent: Interventions in contemporary political debate*. Bloomington, IN: Indiana University Press.

Deutsche Welle. (2016, November 6). Anti-drone rally at US air base Ramstein draws thousands. *Deutsche Welle*. Retrieved from http://www.dw.com/en/anti-drone-rally-at-us-air-base-ramstein-draws-thousands/a-19323459

Gamson, W. A., & Wolfsfeld, G. (1993). Movements and media as interacting systems. *Annals of the American Academy of Political & Social Science*, 528, 114–126.

Gitlin, T. (1980). *The whole world is watching: Mass media and the making and unmaking of the new left*. Berkeley, CA: University of California Press.

Goodman, A. (2016, January 19). Photographing drone war protests lands peace activist, grandmother 6 months in prison. *Democracy Now!* Retrieved from https://www.democracynow.org/2016/1/19/photographing_drone_war_protest_lands_peace

Hanrahan, J. (2011, July 22). At least 1400 arrests for antiwar dissent, but who's counting? Not the press. *Nieman Watchdog*. Retrieved from http://www.niemanwatchdog.org/index.cfm?fuseaction=background.view&backgroundid=563

Harris, P. (2013, March 27). Anti-drones activists plan month of protest over Obama's "kill" policy. *The Guardian*. Retrieved from https://www.theguardian.com/world/2013/mar/27/anti-drone-activists-protest-obama

Holpuch, A. (2015, April 1). Graphic anti-drone video ad to run on television near US air force bases. *The Guardian*. Retrieved from https://www.theguardian.com/world/2015/apr/01/anti-drone-television-ad-us-air-force-bases-california-nevada

Inside Out. (2014, April). Not a bug splat. *Inside Out Project*. Retrieved from http://www.insideoutproject.net/en/group-actions/pakistan-undisclosed-location

Kelly, K. (2016, July 7). Of lethal drones and police shootings. *Consortimnews.com*. Retrieved from https://consortiumnews.com/2016/07/07/of-lethal-drones-and-police-shootings/

KnowDrones Tour. (2012, April 5). National drone tour starts in Brooklyn on April 12, 2012. *PR NewsWire*. Retrieved from http://www.prnewswire.com

Kupfer, D. (2016, April 7). Peace activists arrested in anti-drone protests. *The Progressive*. Retrieved from http://progressive.org/dispatches/peace-activists-arrested-anti-drone-protests/

Lazare, S. (2013, November 15). Anti-drone movement speaks: "End the secrecy, no to kill list." *Common Dreams*. Retrieved from http://www.commondreams.org/news/2013/11/15/anti-drone-movement-speaks-end-secrecy-no-kill-list

Marcetic, B. (2016, August). Meet Chris Antal, the US Army chaplain who resigned over the drone program. *In These Times*. Retrieved from http://inthesetimes.com/article/19276/the-army-chaplain-who-said-no-to-drones

Moyers, B. (2016, January 20). A drone protestor heads to jail. *BillMoyers.com*. Retrieved from http://billmoyers.com/story/a-drone-protestor-heads-to-jail/

Mulder, J. T. (2013, April 29). 30 protesters arrested at Hancock air base at conclusion of weekend rally (names). *Syracuse.com*. Retrieved from http://www.syracuse.com/news/index.ssf/2013/04/30_drone_protestors_arrested_a.html

NBC News. (2014, May 7). Code Pink holds anti-war protest at White House. *NBC News*. Retrieved from http://www.nbcnews.com/watch/nbc-news/code-pink-holds-anti-drone-protest-at-white-house-249217603859

Norton, B. (2016, June 8). US Army minister leaves "imperial" military to protest drone "unaccountable killing," nuclear "policy of terror." *Salon.com*. Retrieved from http://www.salon.com/2016/06/08/u_s_army_minister_leaves_imperial_military_to_protest_drone_unaccountable_killing_nuclear_policy_of_terror

Pater, R. (2013). *Drone survival guide*. Retrieved from http://www.dronesurvivalguide.org

Pratt, M. B. (2014, May 2). Anti-drone events in Central New York. *Workers World*. Retrieved from http://www.workers.org/2014/05/02/anti-drone-events-central-new-york/#.WH-PFRTBDEE

Press TV. (2015, September 27). Germans protest against US drone attacks. *Press TV*. Retrieved from http://www.presstv.com/Detail/2015/09/27/430902/Germany-US-base-Ramstein-killer-drones

Project Censored. (2016, April 4). Peaceful anti-drone activism goes under-reported. *Project Censored*. Retrieved from http://projectcensored.org/peaceful-anti-drone-activism-goes-reported/

Queally, J. (2015, January 5). With call to "end the drone wars," activists cut their way into UK air force base. *Common Dreams*. Retrieved from http://www.commondreams.org/news/2015/01/05/call-end-drone-wars-activists-cut-their-way-uk-air-force-base

Reprieve Foundation. (2014, July 4). Not a bug splat! [Video file]. Retrieved from https://www.youtube.com/watch?v=F0WjpDVDwcM

Ross, A. (2016, July 1). Former US drone technicians speak out against programme in Brussels. *The Guardian*. Retrieved from https://www.theguardian.com/world/2016/jul/01/us-drone-whistleblowers-brussels-european-parliament

Ryan, C. (1991). *Prime time activism: Strategies for grassroots organizing*. Boston, MA: South End Press.

Schogol, J. (2015, April 9). Commercials urge drone pilots to refuse to fly missions. *Air Force Times*. Retrieved from https://www.airforcetimes.com/story/military/2015/04/09/commercials-against-drone-strikes/25532163/

Schonfeld, Z. (2014, April 8). #NotABugSplat, an art project designed to be seen by drones. *Newsweek*. Retrieved from http://www.newsweek.com/notabugsplat-art-project-designed-be-seen-drones-245191

Shesgreen, D. (2015, June 24). McCaskill: Drone warfare taking toll on pilots. *Springfield News-Leader*. Retrieved from http://www.news-leader.com/story/news/politics/2015/06/24/mccaskill-drone-warfare-taking-toll-pilots/29240491/

Shesgreen, D. (2016, April 29). Activists to protest drone attacks. *Springfield News-Leader*. Retrieved from http://www.news-leader.com/story/news/politics/2016/04/29/activists-protest-drone-attacks/83683426/

Solomon, N. (2005). *War made easy: How presidents and pundits keep spinning us to death.* Hoboken, NJ: John Wiley & Sons.

Solomon, W. (2000). More form than substance: Press coverage of the WTO protests in Seattle. *Monthly Review, 52*(1), 12–20.

Swan, R. (2013, November 11). Drone primer now on display in muni bus shelters. *SF Weekly*. Retrieved from http://archives.sfweekly.com/thesnitch/2013/11/11/drone-primer-now-on-display-in-muni-bus-shelters

Weiner, M. (2015, May 27). TV ads urge Syracuse drone pilots to refuse flying over Afghanistan. *Syracuse.com*. Retrieved from http://www.syracuse.com/politics/index.ssf/2015/05/tv_ads_urge_ny_air_national_guard_drone_pilots_to_refuse_flying_over_afghanistan.html

Whitton, M. (2016, May 17). This US Army chaplain just stood up to the American "empire." *AntiMedia*. Retrieved from http://theantimedia.org/this-us-army-chaplain-just-stood-up-to-the-american-empire/

· 11 ·

"I HAVE A DRONE"

Internet Memes and Digital Dissent

While some political memes are framed in a humorous manner, others are deadly serious. But regardless of their emotional keying, political memes are about making a point—participating in a normative debate about how the world should look and the best way to get there.

—Limor Shifman, *Memes in Digital Culture*

In the final chapter of his book of cultural criticism, *Drone*, Adam Rothstein likens these flying machines to a *technological* meme: "They embody fear, paranoia, dreams, and cutting-edge technology as much or more than any other current symbol" (2015, pp. 140–141). Without making explicit reference to drones and the sublime, Rothstein's analysis corresponds with a good deal of my thinking about this new technology. Rather than pursue Rothstein's provocative line of inquiry into the meanings of drones as material artifacts, and in keeping with this project's discursive orientation, this discussion takes a different tack. Specifically, this chapter examines the *internet* meme "I Have a Drone"—an unusually provocative instance of digital culture that calls out Barack Obama on his targeted killing program—with an eye toward exploring the emergence of an online oppositional discourse to drone warfare.

I locate this discussion in relation to contemporary theorizing on internet memes, with an emphasis on the cultural politics enacted within and through this emergent form of "participatory culture" (Jenkins, 2006; Milner, 2012). Throughout, I argue the "I Have a Drone" meme (hereafter IHAD) challenges and disrupts media narratives that unproblematically equate Barack Obama's presidency with the realization of Dr. Martin Luther King Jr.'s "dream"—let alone the slain civil rights leader's legacy of nonviolent resistance to "poverty, racism, and militarism" (King, 1967). As such, IHAD registers and articulates popular moral condemnation of Obama's drone wars and, more generally, cultural anxieties associated with the rise of the drones.

The chapter proceeds with a brief discussion of internet memes and their potential for political expression and engagement. Following this, I review media narratives that liken Barack Obama and Martin Luther King Jr. and suggest that IHAD is best understood as an "ironic intervention" (Day, 2011; Howley, 2013) into the rhetorical equivalence between the two men common to US news coverage of the Obama presidency. As such, this analysis considers the geopolitical significance of memes for the expression of dissident opinion in digital space. I conclude with some thoughts on the limits and possibilities of the prosaic political participation afforded by internet memes.

Internet Memes and Cultural Politics

Coined by evolutionary biologist Richard Dawkins (2006) to capture the virus-like transmission and mutation of cultural ideas and artefacts, the neologism "meme" (a combination of "gene" and "mimesis") has been adopted by internet users to describe the rapid spread, replication, and imitation of digital content. "In the vernacular discourse of netizens, the tag 'internet meme' is commonly applied to describe the propagation of items such as jokes, rumours, videos, and websites from person to person via the internet" (Shifman, 2014, p. 2). Despite the proliferation of internet memes in popular and academic discourse, however, the concept's applicability to the study of digital culture has been challenged (Burman, 2012; Johnson, 2007; Rose, 1998). Jenkins (2009) calls attention to the difficulty of applying biological metaphors to cultural production and transmission. Accordingly, the use of biological or epidemiological analogies to define internet memes reduces netizens to passive recipients (or victims) of mind viruses, rather than active agents in the production of culture. As we shall see, the political dimensions of internet memes runs counter to this argument and instead highlights users' agency in challenging political authorities, media discourses, and dominant ideologies.

To this end, communication scholar Limor Shifman refines Dawkins' initial formulation for purposes of cultural analysis and critique. Contra Dawkins, Shifman understands memes not as single or discrete "units of culture," but rather as a group or collection of texts. Specifically, Shifman defines internet memes as "(a) a group of digital items sharing common characteristics of content, form, and/or stance, which (b) were created with awareness of each other, and (c) were circulated, imitated, and/or transformed via the internet by many users" (2014, p. 41). Drawing on this definition, I submit that by providing "alternative routes of expression for marginalized groups" (Shifman, p. 173) internet memes are emblematic of twenty-first century cultural politics. Put differently, in an era marked by media convergence, networked communication, and the proliferation of user-generated content, memes are part and parcel of "processes through which relations of power are asserted, accepted, contested, or subverted by means of ideas, values, symbols, and daily practices" (Glick Schiller, 1997, p. 2).

For example, in a rhetorical analysis of the "Pepper Spraying Cop" and "Doing a Lynndie" memes, Stefka Hristova observes how two iconic images of state violence—University of California, Davis police officer John Pike pepper spraying members of the Occupy movement, and Lynndie England's infamous "thumbs up" pose with Iraqi detainees at Abu Ghraib—were appropriated, manipulated, and repurposed in ways that "reflected and refracted popular discourses about war, incivility, and political authority" (2013, p. 84). Similarly, IHAD is an especially generative if not altogether well known digital artefact that draws on, remixes, and intervenes in political and popular discourses surrounding the Obama presidency. In short, through digital manipulation, image juxtaposition, and ironic humor, political memes constitute a form of "visual argument" (Hahner, 2013) that artfully blends popular culture and politics in entertaining and incisive fashion.

Media Narratives and Collective Memory

Comparisons between Martin Luther King Jr. and Barack Obama have circulated in news and political discourse ever since then-Illinois State Senator Obama gave a rousing keynote address at the 2004 Democratic National Convention. Obama's soaring rhetoric and charismatic style invariably led pundits to liken the ambitious young senator with the slain civil rights leader. In the intervening years, Obama vacillated between aligning himself with King's oratory and distancing himself from King's legacy. Conversely, news coverage of Obama's presidency was marked by frequent allusions to Dr. King, and specifically, to King's iconic "I Have a Dream" speech. As Eric Sundquist notes, "Obama's election was bound to be perceived as the prophesied fulfilment of the dream 'deeply rooted in the American dream' made famous by King in 1963" (2011, p. 114). Below, I briefly trace the contours of the rhetorical equivalence between King and Obama before moving on to an analysis of IHAD and its ironic deployment of the dream trope.

Living the Dream

Barack Obama is certainly not the first politician to reference Martin Luther King's "I Have a Dream" speech of August 28, 1963. However, as Sundquist notes, "both in the words he spoke and in the very fact of his candidacy" (p. 114), Obama invites comparisons with Dr. King like few others. Over the course of the 2008 presidential campaign, Obama frequently referred to himself as a beneficiary of King's courage, leadership, and moral vision. In turn, journalists and pundits echoed and amplified Obama's personal, rhetorical, and political affinity with Dr. King. At the same time, Obama's opponents—Democrats and Republicans alike—attempted to link the candidate with more radical elements of the civil rights and black liberation movements. Given the vexed history of race relations in the United States, it could scarcely be otherwise. The uproar surrounding Obama's relationship with Reverend Jeremiah Wright is a case in point (Anon, 2008).

When Obama prevailed on November 4, 2008, becoming the first black president of the United States, the stage was set for the formation of what Kristen Hoerl (2012) calls "a postracial narrative of national transcendence" (p. 178). Hoerl's analysis is relevant here insofar as it highlights the "selective amnesia" at work in media narratives that not only "positioned Barack Obama's election as the realization of a postracial version of Martin Luther King's 'dream' and the culmination of the struggle for civil rights," but also, crucially, obscured and elided King's fierce opposition to American militarism. Significantly, the "suturing" (Hoerl, p. 194) of King's dream to Obama's election was evident in media outlets across the political spectrum. Thus, Hoerl underscores the decisive role media narratives play in constructing collective memory. Equally important, Hoerl's analysis foregrounds the contested character of public memory and highlights its heuristic value for interpreting the present and guiding political action. As we shall see, IHAD is especially productive in overcoming selective amnesia and putting collective memory to contemporary political use.

A Dream Deferred?

If equating King's dream with Obama's election resulted in "the negation of the ideals and strategies that distinguish them" (Hoerl, p. 187), Obama's presidency soon revealed important distinctions between the two leaders. Scarcely a year into his first term in office, President Obama was awarded the Nobel Peace Prize. When the Nobel committee announced its decision in October 2009, news reports focused on whether or not the newly elected president had earned the award. Comparisons between Obama and King, who won the award in 1964, quickly emerged as prominent tropes in news discourse. For instance, during an October 12th interview with then-Secretary of State Hillary Clinton, NBC's Ann Curry (2009) said: "The president is joining the ranks of Martin Luther King, of Mother Teresa, of Nelson Mandela. Does he deserve to be in those ranks?" While Mandela, Mother Teresa and other Nobel laureates were occasionally referenced in news reports, journalists and pundits routinely invoked Dr. King in their assessment of Obama's merit for the prestigious award.

Speaking on Fox News Channel, historian Douglas Brinkley articulated a connection between Dr. King and President Obama when he suggested the award represents the Nobel committee's support and encouragement for the new president's diplomatic efforts:

> I think you have to really think about it more in the terms of Martin Luther King. He won a Nobel Peace Prize at 35 years old, right in 1964, before the historic Civil Rights legislation of '65. And King was at that point 35. He had his whole career ahead of him. So it was kind of an award to encourage King to continue fighting for Civil Rights. (Van Susteren, 2009)

At the award ceremony in December 2009, Nobel committee chairman, Thorbjørn Jagland, likewise invoked King's memory. Explaining the committee's decision, Jagland claimed Obama's election was "a call to action for us all" (Jagland, 2009). Turning to Mr. Obama, he concluded, "Dr. King's dream has come true" (Zeleny, 2009). In his acceptance speech, Obama also referred to King, and another icon of nonviolent resistance, Mahatma Gandhi. Praising their efforts, Obama (2009) nonetheless drew a distinction between their leadership and his responsibilities as head of state. "I cannot be guided by their examples alone," he said.

While media reaction to Obama's speech was mixed, the figure of Dr. King remained constant. For example, a New York *Times* editorial praised Obama for acknowledging "without Dr. King's vision, leadership and sacrifice, he never would have been standing at that lectern in Oslo" (Anon, 2009). The *Times* visually reinforced the correspondence between Obama and King when it pictured the President in Oslo alongside archival images of Dr. King and another black Nobel laureate, Nelson Mandela.

More critically, press accounts of Obama's acceptance speech lent tacit approval to the President's recent decision to commit an additional 30,000 US troops to Afghanistan. In an endorsement of Obama's "just war" thesis, the Washington *Post* observed:

> While the president invoked the Rev. Martin Luther King Jr., and called himself a "living testimony to the moral force of nonviolence" Obama also recalled the halting of Hitler's army during World War II to argue that, sometimes, only force can resolve injustice and protect civilians. In an echo of predecessor George W. Bush, he noted that "evil does exist in the world." (Fletcher & Wilson, 2009, p. 1)

In sum, press coverage of Obama's acceptance speech invoked Dr. King for rather limited comparison of the two African-American leaders based on race and rhetoric: an assessment that all but ignores King's vehement opposition to American militarism. Doing so, these accounts gloss over King's legacy of nonviolent resistance, not to mention the inherent contradictions of a wartime president claiming a Nobel Prize for peace. Significantly, it is here that the origins of the "reluctant warrior" meme that has come to characterize US press coverage of Obama's war on terror can be detected (Hart, 2014; Parker, 2009).

In contrast, observers in the independent and alternative press challenged mainstream media narratives that emphasized parallels between King and Obama. These dissident voices marked the occasion of Obama's second inaugural on January 21, 2013—as fate would have it, the same day the nation observed the Martin Luther King Holiday—to highlight glaring contradictions between King's activism and Obama's policies. Obama's drone wars were a focal point for much of this criticism.

The Dream @ 50

In August 2013, as the nation prepared to celebrate the 50th anniversary of King's "I Have a Dream" speech, fresh comparisons between President Obama and Martin Luther King Jr. emerged. As he prepared to deliver a commemorative address from the same location on the Lincoln Memorial where Dr. King told the nation of his dream, the President and his staff sought to downplay such comparisons. That same week, President Obama was weighing military options in light of allegations that the Syrian regime had used chemical weapons against civilians in its protracted civil war. Nonetheless, the historic alignment of America's first black president celebrating one of the most iconic speeches in US history all but insured that such comparisons would be unavoidable, regardless of White House spin. Instead, news organizations stitched the two storylines into a coherent narrative of the President confronting two difficult situations simultaneously.

For instance, NPR's *All Things Considered* co-host Melissa Block (2013) told listeners: "This week, President Obama is engaged in two challenges—possible US military action against Syria and how to mark the 50th anniversary of the March on Washington, an iconic moment in the civil rights movement." White House efforts to downplay comparisons between King and Obama during the weeklong ceremonies constituted a major plot point in the unfolding narrative. The confluence of news coverage, political rhetoric, and national commemoration reinforced the association between Dr. King's legacy and Obama's presidency, further burnishing Obama's reluctant warrior image. As the New York *Times* noted:

> For Mr. Obama, Dr. King has been an idol, role model and burden since he assumed the presidency. He keeps a bust of the civil rights leader in the Oval Office along with a framed program from the 1963 march, and some of his favorite lines have been adopted from Dr. King. But as the nation's first black president, Mr. Obama has found that no matter how much supporters may want to compare them, he cannot be a latter-day Dr. King. (Baker, 2013, p. A14)

Cable and broadcast television likewise focused on similarities and differences between the two men. CNN's Wolf Blitzer (2013) observed how fitting it was for Obama to honor Dr. King for "an iconic speech that paved the way" for his presidency. Thus, the "split-screen comparisons" (Shear, 2013) between King and Obama common to the presidential inaugurations were all the more prevalent in the days leading up to the 50th anniversary of King's speech. And, just as it was on previous occasions, news outlets from across the political spectrum linked the two black leaders visually and rhetorically. For instance, Fox News Channel edited archival footage of King with contemporary footage of Obama speaking from the Lincoln Memorial (Anon, 2013a). ABC's *Nightline* (Anon, 2013b) employed the same visual strategy in its coverage of the President's address:

> KING: We have also come to this hallowed spot to remind America of the fierce urgency of now. Now is the time to make real the promises of democracy.
>
> OBAMA: We may not face the same dangers of 1963, but the fierce urgency of now remains.

Linking Obama with King through words and images was not the sole purview of TV news organizations. As the Washington *Post* noted, "many of the buttons, T-shirts and posters for sale at the event featured images of the two men side by side" (Milbank, 2013, p. A2). As we shall see, IHAD employs this side-by-side visual motif to make ironic comment on media discourse that equates the two African-American leaders.

In sum, since his arrival on the national stage, Barack Obama's rhetoric, and the media discourse surrounding his campaign and presidency, have frequently taken up implicit and explicit comparisons between Dr. King and Mr. Obama. None of which is to suggest that the comparisons were always favourable (e.g., Smiley, 2013). Nevertheless, as Hoerl (2012) notes, critical voices were routinely ignored or marginalized in mainstream media narratives. It is in this discursive context that this chapter analyzes IHAD in an effort to better understand how participatory culture in general, and internet memes in particular, enable non-traditional actors to "express their political opinions, engage in heated debates, and enjoy the process to boot" (Shifman, p. 144).

Drone Memetics

On January 16, 2013, *Common Dreams*, a progressive news site, published an essay by author and activist Norman Solomon (2013) titled "King: I Have a Dream. Obama: I Have a Drone." Pegged to the coincidence of Obama's second inauguration and the King national holiday, Solomon's critique revisits Obama's Nobel Prize acceptance speech to highlight the profound discrepancy between King's nonviolence and Obama's drone wars. Citing a report in *The Guardian* claiming the Obama administration underreports civilian casualties of US drone strikes (Hopkins, 2013), Solomon argues that drone campaigns in Yemen, Afghanistan, Pakistan, and Somalia violate international law and threaten global security. Noting that over 50,000 Americans signed a petition calling for a ban on weaponized drones, Solomon concludes: "Count on lofty rhetoric from the Inaugural podium. The spirit of Dr. King will be elsewhere."

The accompanying uncredited illustration features two images: one, an archival photograph of Dr. King speaking at the 1963 march on Washington; the other, a black and white photo of Barack Obama on the campaign trail. The two men strike remarkably similar poses. However, the explicit correspondence between the two leaders is undercut by text that echoes Solomon's headline. Eye catching to be sure, the illustration does more than provide a visual supplement to Solomon's analysis. Rather, it distills Solomon's argument with the humor, brevity, and insight that are hallmarks of the visual rhetoric of internet memes.

Drawing on Shifman's framework for interpreting the content, form, and stance of internet memes, I examine memes that employ a "split screen" or "side-by-side" visual arrangement to (dis)articulate the relationship between King and Obama. Following this, I consider creative derivatives that abandon the side-by-side comparison, and instead place the figure of Obama, and his drones, in various historical and pop culture contexts. As we shall see, the IHAD meme and its derivatives illustrate the emergence of an oppositional drone discourse in digital space.

Split Screens and Moral Failings

One of the animating forces behind political memes is the reappropriation of news discourse and popular culture to critique, resist, or challenge political and symbolic authority. The meme-based political discourse taken up in IHAD is especially eloquent, not only in its sly critique of media narratives that suture King's legacy with Obama's presidency, but also in its fiercely ironic stance toward Obama's drone wars. To these ends, IHAD frequently employs symmetrical images wherein the two men occupy the same space in the frame, use similar gestures, and wear the same facial expressions. Doing so, the split screen and side-by-side motifs imitate and exaggerate the rhetorical and visual tropes common to the media narratives discussed above.

In this IHAD iteration, the top image from the 1963 march on Washington features Dr. King speaking before the crowd, his mouth open wide and his arm raised high. The bottom image is a contemporary photograph of Obama. Like King, he is surrounded by supporters and, with his arm in the air, addresses an enthusiastic audience. The use of a black and white image of Obama mimics and reinforces the correspondence between the two men frequently employed in news discourse. Moreover, the black and white image echoes and amplifies the "civil rights nostalgia" Hoerl (2012, p. 183) detects in press coverage of Obama's first inauguration.

The text below each image further accentuates the correspondence between the two men. However, unlike political memes that employ image juxtaposition to get a laugh, and sometimes to make a point, ironic word play on Dr. King's refrain signals the incongruity between the two African-American leaders. Put differently, the turn of phrase, "I have a drone" operates as a "countermeme" (Shifman, p. 130) to the celebrated mantra of the civil rights movement. In this respect, IHAD recalls the use of countermemes in subvertisements—advertising parodies that upend and repurpose corporate logos and slogans in support of an anti-consumerist ideology (Harold, 2007). Likewise, the countermeme "I have a drone" subverts the rhetorical equivalence between King and Obama. In this way, IHAD upsets media narratives that too easily conflate the two men based on racial identity and rhetorical gifts, and instead invites viewers to question Obama's drone wars in light of Dr. King's philosophy of nonviolence.

Retaining the split screen motif, IHAD derivatives modify the two images in subtle but productive fashion. For instance, an edited image pairs pictures of King and Obama, both figures gesturing with their hands. Reading left to right, a smiling Dr. King waves to a crowd of supporters. Opposite, the lone figure of Barack Obama wears a stern expression, aggressively pointing his finger. Again, the images are quite similar in composition and framing. However, in this iteration, the two figures literally confront one another. The background and text further distinguishes King from Obama. Here, King's refrain is rendered in black type against a white background; Obama's rejoinder uses white text against a black background. As in the previous example, this collage employs a visual rhyming scheme that features slight, but significant differences to upset the rhetorical equivalence between the preacher and the politician.

Yet another variation juxtaposes a sepia-toned image of a stern Dr. King with a color image of a laughing Obama. Here the intertextual quality of IHAD as it propagates and mutates can be detected—one of the defining characteristics of internet memes Shifman identifies. In addition to reversing the smiling King/aggressive Obama motif, this iteration self-consciously references the better known, and occasionally racist, "Smiling Obama" and "Laughing Obama" memes. This motif of a smiling or laughing Obama is especially common to the creative derivatives discussed below.

While IHAD's split screen iterations offer ironic comment on the discursive correspondence between King and Obama, side-by-side derivatives double down on this impulse by placing the two leaders in the same historical moment. As Hristova notes, "Political meanings are constructed through interplay of the past and the present" (p. 88) and, indeed, this interplay is evident in the media narratives discussed above. But as noted earlier, conflating

King's dream with Obama's election obscures the complex history of the civil rights movement and, more specifically, ignores Dr. King's vehement opposition to American militarism. In contrast, memes that put King and Obama in a single image undermine the equivalence common to media narratives by placing the two men, and by inference, their opposing political philosophies, in the same (time) frame.

Significantly, these IHAD variations pick up on a leitmotif in press coverage of the Obama presidency; namely, the "what if" scenario, wherein Mr. Obama meets Dr. King. For instance, on the occasion of Obama's second inauguration, journalist and activist Tavis Smiley (2013, p. B1) wrote, "I have often wondered what the bust of King in the Oval Office would whisper to the president when he's working alone late at night." One such iteration has the President sitting alongside Dr. King. Again, the two men assume similar poses; both look into the distance and wear troubled expressions, each resting his head in his hand. No less ironic than our previous examples, the corresponding text nonetheless gives the impression of a morally fraught dialogue between the two men. This "exchange" between a champion of nonviolence and a wartime president underscores the discrepancy between each man's political philosophy and practice—a profound difference routinely downplayed in press accounts. Another version turns this meeting into direct confrontation. Here, King and Obama are seated across from one another. As with the previous examples, the framing and positioning of each man is quite similar. But despite their respective poses, the two leaders clearly do not see eye-to-eye. A slight, but meaningful change in tense—the text above King reads, "I *Had* a Dream …"—signals Dr. King's disappointment with America's first black president, further emphasizing the discrepancy between King's dream and Obama's drones (Figure 11.1).

As Hristova notes, iconic images, like the ones routinely employed in news reports—and subsequently repurposed and remixed in meme-based political discourse—serve as "visual shortcuts to both moral judgments and, more broadly, to complex narratives about times past that bear upon the present" (p. 93). IHAD variations that offer split screen or side-by-side comparisons between Dr. King and President Obama draw on a deep reservoir of rhetoric and imagery from the civil rights era. Whereas news discourse typically deploys these words and images to depoliticize Dr. King's legacy, IHAD memes featuring King's likeness privilege questions of morality and the rule of law that are all too infrequent in news and official discourse surrounding Obama's drone wars.

Figure 11.1: Instance of the "I Have a Drone" meme depicting a meeting between Martin Luther King, Jr. and Barack Obama. *Source: quickmeme.com/meme/3snodk*.

Days of Future Past

The tagline of William Banzai's Twitter profile succinctly captures the graphic artist's cultural politics: "Destroy the image and you will break the enemy" (Banzai, 2015). In July 2012, Banzai produced an early iteration of IHAD. In the interim, Banzai continued to produce provocative art that subverts Obama's reluctant warrior image. Unlike the illustrations discussed above, however, Banzai's collage does not incorporate Dr. King's likeness. Nonetheless, the photomontage deploys visual history to invoke the civil rights leader, thereby activating collective memory to critique Obama's drone wars. In this way, Banzai destabilizes Obama's public image: an image constructed, in large part, by news and political discourse that likens Obama with King.

Remixing an archival photograph from the 1963 march on Washington, Banzai crudely inserts Obama's head where King's should be. In the figure of Dr. King, a stern Obama addresses the crowd. In the original photograph, King raises his arm in exhortation to the assembled masses to recommit themselves to the nation's founding principles in the struggle for racial justice. Obama's gesture takes on a more sinister meaning with the insertion of a single drone flying ominously above the crowd. The countermeme, "I have a drone" appears in the upper right-hand corner of the image. Here again, an iconic image from the civil rights era is repurposed to offer critical commentary on the present: a moment that finds America's first black president, a self-proclaimed beneficiary of Dr. King's sacrifice and moral leadership, waging war by remote control across the globe.

Banzai returns to the past to comment on the present in another collage that remixes iconic photographs with contemporary visuals. However, rather than draw on the rich visual history of the civil rights era, Banzai evokes the Vietnam War to critique Obama's drone campaigns. Here, Banzai repurposes a familiar and disturbing image of Vietnamese children fleeing napalm bombs by inserting drones flying overhead. In the middle distance, a smiling Obama appears, not unlike the Baby Sun character of *Teletubbies* fame, rising from a black cloud. Horrifying and darkly comic, this IHAD variation has no text, save for the artist's signature (Figure 11.2). Nonetheless, the image makes a powerful visual argument, one that echoes Dr. King's less well known,

Figure 11.2: "I Have a Drone" creative derivative featuring Vietnam era drone strike with smiling Obama. *Credit: William Banzai. Source: WilliamBanzai7.*

but equally forceful "Beyond Vietnam" speech: an address King made on April 4, 1967, a year to the day before he was assassinated (King). In that speech, the Nobel laureate made an observation that enraged critics and allies alike. Equally important, for my purposes here, it is a bold political statement that has since been censored in ritual commemorations of Dr. King. Without the benefit of direct reference to King, the visual argument advanced in this collage is crystal clear: as it was during the Vietnam era, in the age of Obama "America is the greatest purveyor of violence in the world today" (King).

While the previous examples reference Dr. King's "I Have a Dream" speech to denounce Obama's drone wars, a final example deploys King's words to decry the prospect of drones operating in domestic airspace. Produced by the now-defunct website FightTyranny.org, the illustration coincided with the President's signing into law HR 658, better known as the FAA Reauthorization and Reform Act of 2012 (US Congress, 2012). Among other requirements, HR 658 authorizes the Federal Aviation Administration (FAA) to expedite processes that would permit local, state, and federal government agencies to operate drones in domestic airspace. Unlike most IHAD variations, the meme alludes to King, not through archival imagery, but rather in an extended parody of the text of Dr. King's dream speech.

Below a headshot of Obama, flanked by a drone, the text reads: "I have a drone that one day will exalt every valley, and every hill and mountain will be made low, the rough places will be made plain, and the restless voices will be silenced." Articulating cultural anxieties associated with domestic drone use, from privacy concerns to fears of government repression, the text concludes with an ominous invocation of King's dream speech: "Safe at last, safe at last, you may have no freedom, but you'll be safe at last." While this take on IHAD is not nearly as succinct as our previous examples, this iteration is unique insofar as it refashions the *text* of King's speech to imagine a dystopian America of the not-to-distant future. As such, this example employs the countermeme "I have a drone" in superbly ironic fashion to express fears surrounding the emergence of a national security state policed by drones.

Terror Drones

A final class of IHAD derivatives remixes elements of political discourse and popular culture. Less serious, perhaps, than the examples cited above, the following examples nonetheless illustrate key features of internet memes and participatory culture. First, IHAD memes that reference aspects of popular culture, from hit TV shows to consumer goods, embody the "evolutionary impulse" (Hahner, 2013 p. 163) of internet memes. That is, the countermeme "I have a drone" survives because users attach it to other discourses. Second, the proliferation of IHAD derivatives and affiliated memes that reference Obama's drone wars reveal a high degree of user engagement and enjoyment. Finally, the intertextual affordances of internet memes reveal the "polysemic potential" (Shifman, p. 150) of visual culture in digital space.

Consider, for example, one of Banzai's later iterations on IHAD. While preserving its withering critique of Obama's drone wars, this derivative draws on a far more prolific internet meme that likens Obama to comic book superhero Batman's nemesis: Joker. Obama is not the only politician whose likeness has been remixed with Heath Ledger's image from *The Dark Knight*, but creative derivatives that merge Obama with Joker are plentiful. Banzai's 2014 take on IHAD retains the notion of a "dangerous clown" but instead of visually referencing Joker, Obama is likened to American television's *Bozo the Clown*.

The collage retains visual elements of Banzai's previous iteration, most notably, a reference to the 1963 march on Washington. However, two drones fly in a darkened sky above the crowd, in the distance, a bolt of lightning strikes the Washington Monument. What's more, Obama as Bozo does not address the crowd, but instead gestures threateningly at the viewer. Doing so, this IHAD iteration expresses popular anxieties surrounding the use of drones in domestic airspace. And yet, in common usage, to call someone a "bozo" is to suggest they are bumbling and incompetent (the epithet "Obozo" is common among Obama's conservative critics). Linking Obama with a beloved, but inept TV clown in this fashion invites multiple readings. While one person may view Obama as buffoon, another may see him as homicidal psychopath. Regardless, Banzai's 2014 version of IHAD not only demonstrates the evolutionary qualities and intertextual dimensions of internet memes, but also indicates the emergence of new forms of political engagement in digital space.

Consider, for example, an episode that illuminates what I call the "dialectics" of hashtag activism. On May 7, 2014, Michelle Obama tweeted a photo of herself holding a sign reading, #BringBackOurGirls. Doing so, the First Lady lent her position and authority to a global campaign calling for the safe return of 276 abducted Nigerian schoolgirls. However, Mrs. Obama's intervention was not universally well received and inadvertently fuelled an anti-drone campaign. Posters took to Twitter and other social media platforms to respond to Mrs. Obama. Some altered Mrs. Obama image with alternative messages. One derivative reads, #MyHusbandKillsKidsWithDrones while another reads simply, #StopDroneMurder. In yet another variation, the image of a dead Muslim girl, presumably a casualty of a US drone strike, includes the text #BringBackYourDrones.

One final example vividly illustrates the intersection of popular culture and politics manifest by internet memes. Titled "Game of Drones," this collage features President Obama on a throne made of assault rifles. A fleet of drones flies over a war torn landscape, while a laughing Obama sits with his finger poised over an ominous red button. There is no reference to Dr. King or his dream. Nonetheless, the collage makes clear, albeit hyperbolic reference to the IHAD memetics discussed above. Moreover, the "Laughing Obama" meme is remixed here with stark imagery and clever word play, thereby inviting multiple interpretations. Drawing on the popularity of HBO's fantasy drama of civil war and political intrigue, *Game of Thrones*, "Game of Drones" is a deceptively complex and ambiguous text: one which operates at the intersection of politics and popular culture in productive and provocative fashion.

***

Limor Shifman observes "meme-based political discourse often begins with a single 'memetic photo' that relates to political actors and controversies" (p. 138). Such is *not* the case with IHAD. The source material for this meme is Dr. King's iconic refrain: "I have a dream." The memetic character of this slogan should not be underestimated. It is the moral clarity of this pre-internet meme that has secured its place in American social and political history. And despite the selective amnesia of media narratives discussed above, King's dream continues to inspire advocacy and activism for social justice. Equally important, King's refrain underscores an important dimension of memes: their role in cultural and political globalization. Indeed, 50 years on, King's dream resonates with people the world over engaged in non-violent resistance to forces of racial, economic, and military oppression.

While the countermeme "I have a drone" may not have mobilized mass demonstrations against the US targeted killing program, it has been a fixture at anti-war protests across the globe. For instance, the New York City Light Brigade, a direct action group that "shines a light on social justice issues," travelled to Washington DC to celebrate the 50th anniversary of King's speech by simultaneously "occupying" the King memorial and the White House with illuminated derivatives of IHAD. And in the wake of Edward Snowden's revelations of NSA spying, thousands marched in Berlin, Germany to protest American militarism and global surveillance. IHAD derivatives were part of the visual fabric of those demonstrations.

As we have seen, IHAD embodies "bottom-up expression" that challenges the uncritical correspondence between Dr. King and President Obama in news and official discourse. Furthermore, IHAD's ironic intervention into this discursive formation articulates profound scepticism toward Obama's reluctant warrior image. More critically, through its evocation of Dr. King, IHAD opens up discursive space to foreground questions of morality and the rule of law in the debate over Obama's drone wars. Put differently, in the absence of robust public deliberation over the legal, ethical, and moral implications of drone warfare, IHAD breaks the silence in an altogether engaging and provocative fashion. Doing so, IHAD exemplifies the capacity of internet memes to "blend pop culture, politics, and participation in unexpected ways" (Shifman, p. 4).

In sum, internet memes enable and encourage non-traditional actors to "speak back" to political authorities in surprising and surprisingly eloquent ways. Equally important, political engagement of this sort underscores the shortcomings of movement politics based on charismatic leadership. That is to say, Dr. King's moral authority is routinely glossed over, especially when it comes to issues of war and peace. So too, Obama's drone war demands skepticism toward any moralizing on his part. In the context of cultural and political globalization, what the IHAD meme demonstrates is the clarity, if not the potential, of a politics based on popular moral leadership.

References

Anon. (2008, April 30). Mr. Obama and Rev. Wright. *New York Times*, p. A18.

Anon. (2009, December 11). President Obama in Oslo. *New York Times*, p. A42.

Anon. (2013a, August 23). Today marks 50 years since MLK's 'I have a dream' speech. *The Five*, Fox News Channel.

Anon. (2013b, August 28). Forward march: The dream goes on. *Nightline*, ABC.

Baker, P. (2013, August 28). President, not preacher, but speaking more about race. *New York Times*, p. A14.

Banzai, W. (2015). Williambanzai7. *Twitter*. Retrieved from https://twitter.com/williambanzai7.

Blitzer, W. (2013, August 23). Obama honors Dr. King's legacy. *The Situation Room*, CNN.

Block, M. (2013, August 27). Middle East, civil rights intersect at White House. *All Things Considered*. Retrieved from http://www.npr.org/templates/story/story.php?storyId=216191873

Burman, J. T. (2012). The misunderstanding of memes: Biography of an unscientific object, 1976–1999. *Perspectives on Science*, *20*(1), 75–104.

Curry, A. (2009, October 12). Interview with Secretary of State Hillary Clinton. *Today Show*. NBC.

Dawkins, R. (2006). *The selfish gene: 30th anniversary edition*. Oxford: Oxford University Press.

Day, A. (2011). *Satire & dissent: Interventions in contemporary political debate*. Bloomington, IN: Indiana University Press.

Fletcher, M. A., & Wilson, S. (2009, December 11). Citing "just war," Obama accepts peace prize. *Washington Post*, p. A1.

Glick Schiller, N. (1997). Cultural politics and the politics of culture. *Identities*, *4*(1), 1–7.

Hahner, L. A. (2013). The riot kiss: Framing memes as visual argument. *Argumentation and Advocacy*, *49*(3), 151–166.

Harold, C. (2007). *Our space: Resisting the corporate control of culture*. Minneapolis, MN: University of Minnesota Press.

Hart, P. (2014, September 26). "Reluctant warrior" Obama bombs yet another country. *AlterNet*. Retrieved from http://www.alternet.org/media/reluctant-warrior-obama-bombs-yet-another-country

Hoerl, K. (2012). Selective amnesia and racial transcendence in news coverage of President Obama's inauguration. *Quarterly Journal of Speech*, *98*(2), 178–202.

Hopkins, N. (2013, January 7). US drone attacks "counter-productive," former Obama security adviser claims. *The Guardian*. Retrieved from http://www.theguardian.com/world/2013/jan/07/obama-adviser-criticises-drone-policy

Howley, K. (Ed.). (2013). *Media interventions*. New York, NY: Peter Lang.

Hristova, S. (2013). Occupy Wall Street meets occupy Iraq: On remembering and forgetting in the digital age. *Radical History Review*, *117*, 83–97.

Jagland, T. (2009). Award ceremony speech. *Nobel Prize*. Retrieved from https://www.nobelprize.org/nobel_prizes/peace/laureates/2009/presentation-speech.html

Jenkins, H. (2006). *Fans, bloggers, and gamers: Media consumers in a digital age*. New York, NY: New York University Press.

Jenkins, H. (2009). *If it doesn't spread, it's dead (part one): Media viruses and memes*. Retrieved from http://henryjenkins.org/2009/02/if_it_doesnt_spread_its_dead_p.html

Johnson, D. (2007). Mapping the meme: A geographical approach to materialist rhetoric. *Communication and Critical/Cultural Studies*, *4*(1), 27–50.

King, M. L., Jr. (1967, April 4). Beyond Vietnam. *The King Institute*. Retrieved from http://kingencyclopedia.stanford.edu/encyclopedia/documentsentry/doc_beyond_vietnam

Milbank, D. (2013, August 29). 2013 didn't live up to 1963. *Washington Post*, p. A2.

Milner, R. (2012). *The world made meme: Discourse and identity in participatory media* (Doctoral dissertation). University of Kansas.

Obama, B. (2009, December 10). Remarks by the president at the acceptance of the Nobel peace prize. *The White House*. Retrieved from http://www.whitehouse.gov/the-press-office/remarks-president-acceptance-nobel-peace-prize

Parker, K. (2009, April 11). An American triumph in Oslo. *Washington Post*, p. A31.
Rose, N. (1998). Controversies in meme theory. *Journal of Memetics—Evolutionary Models of Information Transmission, 2*(1). Retrieved from http://cfpm.org/jom-emit/1998/vol2/rose_n.html
Rothstein, A. (2015). *Drone*. New York, NY: Bloomsbury Publishing.
Shear, M. D. (2013, April 23). Shaping a speech, 50 years after "I have a dream." *New York Times*, p. A16.
Shifman, L. (2014). *Memes in digital culture*. Cambridge, MA: MIT Press.
Smiley, T. (2013, August 25). Obama must be King-like, not King-lite. *Washington Post*, p. B1.
Solomon, N. (2013, January 16). King: "I have a dream." Obama: "I have a drone." *Common Dreams*. Retrieved from http://www.commondreams.org/views/2013/01/16/king-i-have-dream-obama-i-have-drone
Sundquist, E. J. (2011). "We dreamed a dream": Ralph Ellison, Martin Luther King, Jr. and Barack Obama. *Daedalus, 140*(1), 108–124.
van Susteren. (2009, October 9). *On The Record with Greta Van Susteren*. Fox News Channel.
United States Congress. (2012, February 14). *FAA Modernization and Reform Act of 2012*. Retrieved from https://www.congress.gov/112/plaws/publ95/PLAW-112publ95.pdf
Zeleny, J. (2009, December 10). Accepting peace prize, Obama offers "hard truth." *New York Times*. Retrieved from http://www.nytimes.com/2009/12/11/world/europe/11prexy.html

·12·

THINK LOCALLY, BOMB GLOBALLY

Satirizing Drones

For your race, in its poverty, has unquestionably one really effective weapon—laughter.
—Mark Twain, *The Mysterious Stranger*

Larry Wilmore, host of the satirical news program *The Nightly Show*, didn't pull any punches during his speech at the 2016 White House Correspondents Dinner. Like his predecessor Stephen Colbert—whose monologue at the same event a decade earlier established the preeminence of parodic news in twenty-first century public discourse—Wilmore took aim at the political and media establishment, including his host, Barack Obama (Landy, 2016). Turning to the President, Wilmore congratulated Obama for enjoying his final year in office. "Saw you hanging out with NBA players like Steph Curry ... You know it kinda makes sense, too, because both of you like raining down bombs on people from long distances." Wilmore's sly reference to Obama's drone wars drew audible gasps from attendees. But the audience's reaction says less about Wilmore's lack of deference to presidential authority than it does the double standards of Beltway elites. Indeed, while the assembled journalists, politicians, and celebrities groaned at Wilmore's quip, President Obama had the same audience in stitches when, at the 2010 dinner, he warned pop music stars, the Jonas Brothers, to keep away from his daughters. "Sasha and Malia are huge fans, but boys, don't get any ideas. I have two words for you: Predator drones. You will never see it coming."[1]

At the time, news organizations questioned the President's taste, but few noted the cruel irony of Obama's joke (Schwarz, 2016). Hours earlier, Faisal Shahzad, a Pakistani-American, attempted to set off a car bomb in New York City's Time Square. In a dramatic instance of the hazards—realized or not—of blowback to America's targeted killing program, Shahzad told investigators he was motivated by US drone strikes across the Muslim world. More to the point, the lighthearted response to Obama's joke offers a telling counterpoint to liberal condemnation of President George W. Bush's attempted humor during the 2004 Radio and Television Correspondents Dinner. At that black-tie event, Bush presented a slide show with photos of him searching the Oval Office—ostensibly looking for Iraq's weapons of mass destruction (Teather, 2004). Recalling Bush's repugnant jest in light of Obama's equally callous wisecrack, Philadelphia *Daily News* columnist Will Bunch tweeted, "Let's be honest, fellow progressives; we'd be all over Bush if he made the same 'predator drone' joke" (Quoted in Edwards, 2010). Bunch's critique of liberal consent to Obama's drone wars neatly encapsulates a prominent discursive theme of America's war on terror; namely, the continuities and contradictions inherent in US counterterrorism strategy under both Bush and Obama.

This chapter explores the intersection between humor, politics, and dissent in contemporary American culture. Specifically, I focus on the use of irony and satire in what I have been referring to throughout as drone discourse. This discussion proceeds with an examination of political cartoons that make implicit and explicit comparisons between Bush and Obama: a mimetic trope, commonly known as "Obusha," that conflates the two men and their respective approaches to the war on terror. As we shall see, these cartoons reflect and refract debates over enhanced interrogation and indefinite detention under Bush, as well as Obama's expansion and escalation of his predecessor's drone program (Figure 12.1). Next, I consider cartoons that integrate news and official discourse on targeted killing with public debate over a number of contentious social issues. My analytical focus on editorial cartoons draws on scholarship that understands the genre as "a form of satirical journalism and a type of visual opinion news discourse" (Greenberg, 2002, p. 182), which performs significant but frequently overlooked cultural work in the construction and deliberation of social problems. Cartoons that "send up" drones are relevant to my concerns here insofar as they integrate, synthesize,

Figure 12.1: Editorial cartoon critiquing the expansion of George W. Bush's drone program under Nobel laureate Barack Obama. *Credit: Nate Beeler. Source: Cagle Cartoons.*

and recirculate a variety of discourses, from politics and commerce to popular culture, and all with equal measures of clarity, concision, and critique.

Following this, I turn my attention to satirical drone discourse in *The Colbert Report*. Like editorial cartoonists, late night comedians invariably draw on current events to get a laugh. As a result, all manner of drones, from pizza delivery vehicles to hi-tech weapon platforms, are the subject of comedy sketches and one-liners. Rather than attempt a comprehensive analysis of this material, I instead focus on Stephen Colbert's ingenuity for challenging news and official discourse surrounding the drone program. My emphasis on Colbert reflects two lines of thought. First, as Amber Day (2011) argues, the parodic news genre "has moved center stage as it has increasingly blurred the line between news and entertainment" (p. 57). Moreover, "discursively integrated texts" (Baym, 2005) such as *The Daily Show*, *The Colbert Report*, and, more recently, *Last Week Tonight* demonstrate the discursive power of irony and satire for political critique and civic engagement (Gray, Jones, & Thompson, 2009). Second, Colbert's reputation for critically interrogating the use and abuse of language in American political discourse is especially well suited to my aims here. To that end, I highlight *The Colbert Report's* perceptive and slyly subversive engagement with drone discourse. Throughout, I hope to demonstrate the political value and importance of opposition to the targeted killing program in the satiric register.

Obusha

Speaking in October 2002, then-Senator Barack Obama famously declared he wasn't against all wars—just dumb ones. Obama criticized the Bush administration's rush to war in Iraq, claiming:

> An invasion of Iraq without a clear rationale and without strong international support will only fan the flames of the Middle East, and encourage the worst, rather than best, impulses of the Arab world, and strengthen the recruitment arm of al-Qaida.[2]

Throughout his historic 2008 campaign, Obama condemned Bush's ill-conceived intervention to depose Iraq's long-time dictator, and one-time American ally, Saddam Hussein. When he assumed office, in 2009, Obama further distanced himself from his predecessor, promising to close the detention facility at Guantanamo Bay, usher in a new era of transparency in government, and bring the wars in Afghanistan and Iraq to quick and decisive resolutions. Thus, the Obama administration set out to distinguish itself from Bush-era policies, such as indefinite detention, enhanced interrogation, and government secrecy.

As for his predecessor's other legacy—the use of armed drones to hunt down and kill al-Qaeda and Taliban leaders in Afghanistan, Pakistan, and Yemen—Obama was uncharacteristically silent. In fact, his administration did not formally acknowledge the existence of the clandestine program until late in Obama's first term, when counterterrorism adviser John Brennan defended the use of drone strikes against suspected terrorists in an address at the Woodrow Wilson Center (Miller, 2012). Doubling down on Bush's targeted assassination program, Obama's carefully scripted distinctions between himself and Mr. Bush became increasingly difficult to maintain, much less discern. Editorial cartoons reflected public sentiment that increasingly viewed the two men as more alike than not.

Consider, for example, a single-panel cartoon published in February 2013. Bush and Obama are engaged in a heated exchange. Bush wears a cowboy hat, an American flag poised on top. Obama dons a business suit with the flag emblazoned on his shoulder. Like mirror images, the two men angrily point their fingers at each other. Against a red, white, and blue backdrop, their words repeat, filling the space between them: "Drones," shouts Obama; "Torture," exclaims Bush. Despite stark differences in their public personas—politically crafted discursive constructions of the highest order—the cartoon suggests there is little difference between the two men, apart from their weapon of choice.

Conversely, Monte Wolverton's panel foregrounds the differences in each man's mediated image. Bush, whose penchant for malapropisms launched a thousand late-night television jokes, wears the bemused look of a village idiot as he operates a remote control. Obama, depicted as a drone, dutifully receives instructions from his predecessor. Obama's impassive expression recalls Hollywood killing machines like *RoboCop* and *Terminator*. Bush says, "Lookit my new toy. I controlify remotely! I call it the Obama administration." Ironically, the reasoned and articulate Obama acts as an unfeeling, unthinking robot, obeying the commands of his inept and inarticulate pilot. Here we can discern the discursive roots of the Obusha meme—the idea that Obama's time in office does not represent a departure, but rather an extension and expansion of Bush's global war on terror—as it emerged and developed in political cartoons (Figure 12.2).

Figure 12.2: Single-panel cartoon draws unflattering comparisons between Presidents George W. Bush and Barack Obama. *Credit: Monte Wolverton. Source: Cagle Cartoons.*

For instance, an illustration published in the Columbus *Dispatch* features Bush and Obama as the baddies from *Austin Powers*. The two men wear matching grey suits. Bush, his finger pressed to his lip à la Dr. Evil, towers over Obama, who likewise adopts Dr. Evil's amusing affectation. In his other hand, Obama holds a document with a litany of Bush-era abuses: Spying on Americans, Secrecy, Drones, and Executive Power. A pleased Mr. Bush proclaims: "I shall call him mini-me!" Equating two American presidents with *parodies* of villains from the Bond film franchise illustrates the multi-layered discourses at work in editorial cartooning. Like the internet memes discussed in Chapter 11, political cartoons routinely draw from discourses of popular culture to interpret and comment upon news of the day (Conners, 2005, 2007). From a historical perspective, then, political cartoons are a prototype for today's discursively integrated satirical news shows. Put differently, political cartoons have been blurring the lines between news and entertainment long before Jon Stewart and Stephen Colbert arrived on the scene.

One final example of the rhetorical equivalence between Bush and Obama in political cartoons, this time from an internationalist perspective, depicts the two men as housekeepers sweeping dead bodies under either end of a carpet. Bush's apron reads: Rendition Programme. Obama's broom bears the inscription: Drone Programme. Eyeing Mr. Obama warily, President Bush says, "You kept your promise to clean up after me …" Notwithstanding the racial undertones of Obama's depiction as a domestic servant, or the feminization of Bush for that matter, the cartoon suggests the two men are indeed working toward the same ends, albeit using different methods. Standing in the middle ground between this odd couple, a haggard Uncle Sam consoles a distraught Lady Liberty. The insertion of these two mythic figures—visual metaphors for freedom, justice, and other "American values"—heightens a sense of tragedy that pervades the cartoon. Obama's drone wars, an extension of Bush-era counterterrorism strategy, have killed hundreds, perhaps thousands of innocent civilians, whose lives—and especially their deaths—go largely unreported in Western media. Furthermore, in their conduct of the war on terror, both Bush and Obama brazenly violated domestic and international law while skillfully manipulating and frequently intimidating journalists and whistleblowers. Without putting too fine a point on it, then, and notwithstanding the Norwegian Nobel Committee's opinion, Obama did not made a decisive break from Bush-era national security policy. The Obama administration's lasting legacy is the expansion of American empire in the twenty-first century: much in line with the aspirations of Dick Cheney,

Donald Rumsfeld, and Paul Wolfowitz, the architects of George W. Bush's foreign policy.[3]

From the Sublime to the Ridiculous

The varied and contradictory meanings associated with drones—from their preeminence in America's war on terror to their emergent retail, commercial, and recreational applications—provide editorial cartoonists with a wellspring of satirical possibilities: from the sublime to the ridiculous. Consider, for example, an editorial cartoon published by the Augusta *Chronicle* in late 2014. In a box insert in the upper left corner, Barack Obama holds a copy of the recently released Senate Intelligence Committee report on the CIA's torture program. The report's highly redacted findings revealed the scale and scope of the intelligence agency's practices of extraordinary rendition, indefinite detention, and enhanced interrogation during the Bush years. The controversial, and by some accounts partisan report provided Obama and the Democrats with a unique opportunity to score political points by drawing legal and moral distinctions between the current administration and the Bush White House. Echoing the President's own words, a caricatured Obama self-righteously declares, "This is not who we are."

The balance of the panel, however, undermines the administration's long-standing (if unsupported) claims that the escalation of the drone program under Barack Obama somehow represents the moral high ground in America's war on terror. Reading from left to right, a Predator drone fires a Hellfire missile at a suspected terrorist. The plane does not bear the usual military markings, but instead sports the familiar Obama insignia—a recurrent visual trope in cartoon depictions of the US drone program. As a missile streaks toward its target with a FOOM! a disembodied voice from the unmanned plane exclaims, "This is who we are!" The word "this" rendered in bold red letters and underlined for further emphasis. Meanwhile, the terror suspect flies through the air, his hands raised, and shouts, "Hands up! Don't Shoot!" The cartoon's intertextual dexterity and discursive promiscuity is not unique, but rather a defining characteristic of the form (Morris, 1993). In other words, one of the hallmarks of editorial cartoons is their ability to synthesize disparate discourses into concise and frequently comical critical commentary (Greenberg).

Here, the insertion of a campaign logo suggests that Obama literally owns the drone program. Indeed, as we noted in Chapter 3, over the course of his first term in office, targeted assassination emerged as the cornerstone of Obama's counterterrorism strategy, effectively making drones part of his political brand. So much so that during his 2012 reelection campaign, and again in 2016, as he prepared to leave office, Obama made highly publicized efforts to provide greater transparency in the targeted killing program—ostensibly to prevent his successors from (further) abusing these lethal operations.[4] Conversely, the disembodied voice betrays the duplicity of the Obama administration's repeated claims that drone strikes represent a more measured and principled approach to counterterrorism than that of his predecessor.

Most audaciously, however, the terrorist employs the rallying cry of the Black Lives Matter movement: "Hands Up, Don't Shoot." Activists popularized this expression in the wake of the 2014 police shooting of unarmed African American teenager, Michael Brown, in Ferguson, Missouri. Brown's murder ignited nationwide protests and inflamed an already heated debate over race relations in the age of Obama. Inserting this domestic call for racial justice in the context of the US drone program makes a bold and surprising—for some—but revealing connection between American law enforcement's excessive use of force against people of color and Obama's drone wars against tribal Islam (Wall, 2016). Accordingly, domestic race relations—founded upon a long history of structural inequality and state-sanctioned violence against black and brown bodies—parallel the global violence perpetrated by the United States and its Western allies against the peoples of the Greater Middle East. In unexpected yet provocative fashion, the cartoon invokes an incisive analysis of structural racism and systematic oppression here in America and across the globe: a line of thought that can be traced from the nineteenth-century abolitionist struggle to today's Black Lives Matter movement.

The larger point is that editorial cartoonists deploy drones as visual tropes to address a range of social issues that simultaneously vex and define American politics and culture. Take for example Mike Keefe's panel from 2014. Pegged to a series of botched executions by lethal injection, the cartoon's absurd tableau is set in a prison yard. At the center of the scene, a prisoner wearing a standard issue orange jumpsuit sits shackled to a chair. The shadow of a Predator drone engulfs the doomed man. In the middle distance three figures, a reporter flanked by two prison officials, stand behind a sandbag perimeter—ostensibly bearing witness to the execution. The first guard assures the reporter, "No more botched executions. We've switched to drones." "Quick and effective," the second guard adds comfortingly. Drawing on the discourse of anti-drone activists, who routinely characterize Barack Obama as judge, jury, and executioner-in-chief, the cartoon likens death by lethal injection—defended by death penalty advocates as a "humane" form of execution—with the "precision" weapons used for targeted assassination. Juxtaposing these two instances of state-sanctioned murder, the cartoon invites readers to interrogate discourse that legitimates and naturalizes execution at home and abroad.

Similarly, a 2015 editorial cartoon published in the Pittsburgh *Post-Gazette* uses drones as a rhetorical device to intervene in what is arguably the most habitual and vociferous debate in American culture: gun control. In mid-air, a Predator drone meets a handgun. The drone, endowed with reason and beset by moral uncertainty, confesses to the firearm, "I accidently killed two innocent people!" "Slacker!" responds the smug and indifferent handgun. Here, the cartoon employs the perennial trope of the conscious machine to interrogate America's sublime fascination with weapons. The cartoon's polyvocal quality is especially productive insofar as it captures a range of competing discourses that shape and reflect our paradoxical relationship with all manner of killing machines.

For instance, the drone uses plain language—"innocent civilians"—rather than the coolly bureaucratic—"collateral damage"—to express its remorse. As noted in Chapter 6, deceptive and purposefully misleading language in news and official discourse performs the crucial ideological work of legitimating the US drone program and obscuring civilian casualties in its wake. Viewed in this light, the cartoon expresses the ethical and moral dimensions of drone warfare with brevity, irony, and a touch of indignation. Read another way, however, the cartoon underscores the moral relativity of debates over the targeted killing program. That is to say, for gun control advocates, handwringing over the loss innocent lives a world away, while well intentioned perhaps, misses a larger point: gun violence at home demands immediate attention. Conversely, the handgun's bravado echoes the discourse of Second Amendment absolutists, whose mantra, "guns don't kill people, people do"—a questionable premise to be sure—productively subverts the trope of the autonomous killing machine.

Contrasted with the relatively open meanings of this cartoon, another example unambiguously and rather poignantly articulates the relationship between gun violence in America and its targeted killing program. Published in the aftermath of the 2012 massacre at Sandy Hook Elementary School, the single panel cartoon features two young boys sitting in an abstractly rendered tribal Muslim village. The first asks incredulously, "How can twenty children get killed in a school in America?" "Must have been a drone strike," replies the second child. The cruel irony is hard to miss and difficult to ignore: Americans invariably respond to mass shootings with shock, horror, and outrage. All the while, news and official discourse reverberate with words like "tragedy," "innocence," and "senseless violence." But rarely do we acknowledge, nor are we asked to consider, the consequences of a targeted killing program that claims untold numbers of innocent lives. Integrating the discourse of school shootings with that of the drone wars, the cartoon inserts moral clarity into an otherwise anemic debate over the justice, if not the wisdom, of Obama's drone wars.

Finally, Pulitzer Prize-winning artist Mike Luckovich's cartoon from late 2013 features a mash up of discourse surrounding the targeted killing program with Amazon's promotional campaign for Prime Air. Published on the heels of Jeff Bezos's *60 Minutes* appearance plugging the online retailer's planned drone delivery service, the cartoon's first panel depicts three bearded men looking skyward. One figure is clearly labeled "Al-Qaida," while another points to the heavens warning his colleagues, "Drone! Take cover!" In the second panel, an Amazon quadcopter bearing a small package lands before the three men. "Relax," says the third man. "It's delivering my juicer." Here, then, popular understandings of drones in US counterterrorism efforts run headlong into Amazon's promotional discourse. Moreover, the cartoon represents yet another link in the intertextual chain initiated by Amazon's highly publicized—and widely parodied—PR campaign.

Unlike our previous examples, however, this cartoon avoids overt satirical critique of the US drone program in favor of exploring the dissonance between remote control warfare, on the one hand, and retail applications of drone technology on the other. Reveling in the absurdity of terrorists enjoying the benefits of state-of-the-art American consumerism, the cartoon juxtaposes Amazon's technological dreams against public apprehension of these killing machines; cultural attitudes which, as we have seen, are difficult to supplant in the popular imagination. Put differently, the cartoon distills a critical discourse moment wherein an emergent discursive formation on the domestication of drone technology confronts the dominant discourse surrounding the US drone campaign.

Two Birds with One Drone

Like editorial cartoons, satirical news programs intervene in political debates through irony, parody, and sarcasm; mocking political figures and institutions, interrogating contentious social issues, and illuminating cultural contradictions, all in entertaining and frequently incisive fashion. One of the foremost practitioners of the genre, Stephen Colbert's satirical take on drones demonstrates the significance of parodic news in a journalistic culture which, as Norman Solomon (2005) has documented, rarely challenges executive authority and invariably supports American militarism. For instance, in Chapter 3 we noted that a New York *Times* report from 2012 describing the Obama administration's counterterrorism operations—including previously unreleased details of the US drone program—failed to gain much traction in American news outlets. Even the *Times* headline, "Secret 'Kill List' Proves a Test of Obama's Principles and Will," frames the story in terms of the personal implications of the drone program for the President; effectively omitting the far more significant legal, strategic, and ethical consequences of killing by remote control. What's more, despite disturbing revelations about the "kill list" Obama and his national security team use to select targets for assassination, domestic news organizations offered little substantive let alone critical commentary. In an all too frequent turn of events, a late night comedy show, rather than an American news organization, offered a penetrating analysis of the kill list in an installment of *The Colbert Report's* long-running segment, "The Word."

As Geoffrey Baym observes, Stephen Colbert's "ironic performance is continuously double-voiced—his literal language always placed in juxtaposition with its implied meaning" (2009, p. 130). The comedic set up for "The Word" relies precisely on this double-voiced technique, with Colbert addressing the viewer on one side of the screen, and textual information—words and phrases—appearing on the other. This unspoken commentary invariably takes a contrarian position; undermining Colbert's assertions with bad puns and snarky comebacks—a bit like real-time fact checking with an attitude. In this segment, Colbert begins with an acknowledgement: civilian casualties of the drone program threaten America's moral authority. "But," he offers hopefully, "the administration has a solution, and it's the subject of tonight's 'Word'—two birds with one drone."[5] Quoting John Brennan's April 2012 address on the secretive drone program, Colbert says, "We only authorize a particular operation against a specific individual if we have a high degree of confidence that the individual being targeted is indeed the terrorist we are pursuing." As he reads Brennan's statement, Colbert emphasizes specific phrases such as "particular operation" and "specific individual." Doing so, Colbert highlights the dense prevarication in Brennan's statement while underscoring the pretzel logic of official discourse used to justify the targeted killing program. Brennan summarizes his Orwellian equivocations with a decisive claim: "This is a very high bar." "Yes," responds Colbert sarcastically, "It is comforting to know that the administration has set the bar at the high level of killing the right person." {*Higher Standard Than Texas*}. Like one of the cartoons discussed above, this unspoken comment equates state-sanctioned murder by lethal injection with targeted assassination.

Next, a news clip features conservative pundit Charles Krauthammer discussing the use of "baseball cards" in Obama's weekly counterterrorism meetings—euphemistically dubbed "Terror Tuesdays." Colbert quips, "Obama gets baseball cards, with pictures and stats on suspected terrorists. And of course, a stick of stale gum, which he gives to Biden." A photomontage features an official portrait of the President alongside an image of the Vice President blowing a bubble. Of course, vice presidents have been perennial targets of American satirists since the early days of the republic. Nevertheless, this fast and rather facile joke at Joe Biden's expense underscores the perversion of a rather innocuous, somewhat sentimental practice—trading and collecting baseball cards—into a cold and calculated bureaucratic exercise in death by remote control. But Colbert is just getting started.

Over a graphic insert of the *Times* story, Colbert reads the following pull quote: "Eventually, the process must work, because a senior administration official said that the number of civilians killed in drone strikes in Pakistan under Mr. Obama was in the 'single digits.'" Unlike his counterparts in the US press corps, who uncritically echo and amplify such official pronouncements, Colbert is skeptical of the administration's assessment of the precision and accuracy of its drone campaign. "That's impressive," says Colbert of the Hellfire missiles launched by America's unmanned planes. "Because those bombs are huge." {*Not John Carter Huge, But Still!*}. This reference to the highly promoted but poorly received Disney science fiction epic demonstrates the subversive use of popular culture in satirical news programs. A fantasy action adventure film, *John Carter* (2012), tells the story of a time traveling American civil war soldier embroiled in an intergalactic war. Despite the attendant hype and promotion, however, *John Carter* "bombed" at the box office. Science fiction. Futuristic warfare. Box office bombs. Popular culture provides a rich resource to interrogate news and official discourse, and to do so in an engaging and productive fashion that promotes critical reflection and engagement (Jenkins, 2009; Thompson, 2009).

Seizing on one of the more pernicious aspects of the US drone program—an accounting scheme that maintains the fiction that hi-tech targeted assassination avoids civilian casualties—Colbert reveals the administration's rhetorical sleight of hand with yet another pop culture reference. "And the administration has developed a brilliant system for insuring that those building engulfing explosions don't kill non-combatants. They just count all military aged males in a strike zone as combatants." Visually, this last phrase serves as a pull quote to emphasize the administration's specious rationale for its duplicitous accounting methods. "They reason that al-Qaeda is an isolated fanatical group, so anyone with them is probably al-Qaeda. The same way that anyone at Comic-Con is probably a virgin." {*Saving Themselves for Buffy*}. Again, *The Colbert Report's* deft use of popular culture for purposes of critical commentary underscores the absurdity of the administration's claims that so-called "non-combatants"—another bit of obfuscatory rhetoric designed to sanitize the human costs of drone warfare—are not endangered by the "precision" ordinance fired from unmanned planes.

Ironically defending Obama's drone wars, Colbert reassures his audience that, contrary to criticism from legal experts and human rights organizations, "This isn't just the President executing innocent people around the world by fiat." Colbert goes on to describe the twisted logic behind the administration's legal protections for innocent victims of the targeted killing program. "There is an appeals process. The men are considered terrorists unless there is explicit intelligence posthumously proving them innocent." Again, a pull quote serves to emphasize the absurdity of such an appeals process. "In which case," Colbert deadpans, "I assume that there is a legal process that unkills them." {*Ali, Ali Oxen Free!*}. Like war planners and politicians, Colbert perverts both language and logic with this bit of newspeak: "unkill." Doing so, Colbert assumes the naïve attitude of news workers whose deference to political elites blinds them to the legal, moral, and ethical implications of the targeted killing program and the official discourse used to justify this murderous practice.

The segment concludes with a masterstroke of irony—one that skewers legal arguments supporting both the targeted killing and indefinite detention programs. Colbert begins, "Now, best of all, keeping civilian casualties down, by saying the civilians don't really exist, could solve the President's Gitmo problem once and for all. Because, you know what that place is full of?" {*Redacted*}. Here, the double-voiced graphic makes a sly but unmistakable reference to the secrecy shrouding US counterterrorism operations. "It is full of known terrorists. He should shut that place down with some drone strikes." {*Guant no mo'*}. Playing devil's advocate for a moment, Colbert acknowledges the downside of his modest proposal. "Now, of course, that might take out some guards." "But," he continues reassuringly:

> Using the President's own standard, they are military aged males spending time with terrorists. They must be guilty of something. And if we just keep using logic like that, none of us have to feel guilty about anything. And that's the word.

Thus, Colbert ruthlessly undermines the specious reasoning that provides legal justification for the US drone program. Equally important, while Colbert directs his ire at news and official discourse throughout much of this segment, his final assessment squarely identifies and challenges the public's consent to both indefinite detention and targeted killing. As we've seen, humorists employ irony and satire for purposes of critical commentary and moral condemnation. Colbert's flair for such biting satire illustrates one of the defining features of contemporary political humor: its earnest and frequently perceptive engagement with vexing social issues (Day, 2011).

The World's Policeman

Good Catholic that he is, Stephen Colbert is plagued by guilt—as his "solution" to the cascading legal, ethical, and strategic problems posed by the prison camp at Guantanamo demonstrates. But Colbert is no apologist for American empire. In the wake of two damning reports issued in October 2013, Colbert ascended his bully pulpit to launch a vociferous defense of the US drone program.[6] In the manner of former Fox News host, Bill O'Reilly—the alpha and omega of Colbert's parody of conservative punditry—the faux news anchor tells it like it is; leavened, as usual, with an ironic twist. "The United States is the world's policeman—that's why we eat so many donuts. But apparently you cannot police 6 billion people these days without getting a few complaints." A montage of news clips describing published reports by Amnesty International and Human Rights Watch documenting civilian casualties of US drone strikes in Yemen and Pakistan follows, ending with one news reporter declaring, "US drone strikes could amount to war crimes."

Routinely ascribed to rogue-states and other official enemies, the phrase "war crimes" rarely passes the lips of American journalists when describing US military affairs. But Colbert is having none of it. "War crimes? No. Technically we haven't declared war on those countries—they're just crimes." The joke's brevity matches its bite. Shrugging off America's criminal behavior as so much business as usual, Colbert reminds viewers that the atrocities detailed by the two human rights organizations were conducted in undeclared battlefields. Without putting too fine a point on it, Colbert's ironic admission that the US conducts drone strikes where ever and when ever it likes embodies the hubris of a global hegemon. "Besides," Colbert continues, "I think we've made it clear that these drone attacks are done with surgical precision." Here, Colbert dismisses out of hand independent investigations of civilian casualties; preferring instead the altogether comforting, yet unsubstantiated, official claims—dutifully repeated in corporate news outlets—that drone strikes eliminate terrorists while sparing the lives of innocent civilians. Colbert's allusion to the "surgical precision" of US drone strikes sets up an extended joke on this deceptive and purposefully misleading phrase.

Colbert continues, "The best way for a surgeon to remove your appendix is by Hellfire missile. We may have vaporized your liver, kidney and spleen—but your appendix is on the run." Accompanying this line, a photomontage depicts a patient awaiting surgery, an armed Predator aimed at his midsection. While the absurd visual generates an easy laugh, the joke mimics official discourse in productive fashion. Politicians and pundits routinely offer platitudes like, "the terrorists are on the run" to legitimate the expansion of the US drone program across the Middle East and the Horn of Africa. Colbert's absurd extension of the anatomical metaphor challenges the accuracy, let alone the logic, behind such statements. Taking another swipe at official claims, if not the credulity of the White House press corps, Colbert offers further reassurance, noting that President Obama recently declared that prior to ordering a drone strike, there must be "near certainty" that no civilians will be injured or killed. "And that's the kind of almost sure thing I'm willing to bet someone else's life on," Colbert quips. With this, Colbert gets straight to the point: the Obama administration's guarantees may be comforting for domestic audiences, but they fail to stand up to critical scrutiny. All the more reason to keep contradictory information, like reports of civilian casualties in America's drone wars, out of sight and out of mind.

Finally, Colbert appeals to another American value: the almighty dollar. Sounding as if he was speaking before the local Chamber of Commerce, Colbert adds:

> And folks, this has been great for the local Pakistani economy. Because of the constant presence of circling drones, sales of sleeping tablets, anti-depressants and medicine to treat anxiety have soared. As well as pills for erectile dysfunction. Yes, there's

a reason that Cialis couple is sitting outside in bathtubs. That's all that's left after the drone blew up their house.

Here, Colbert's boosterism for American empire aligns with his abiding reverence for consumer capitalism; thereby making explicit the relationship between US militarism on one hand and neoliberal economics on the other. As Norman Solomon observes: "The connections between military might and global commercial market share are not shouted from Washington's rooftops, but the links are solid" (2005, p. 116). More daringly, and at the risk of making light of the psychological terror induced by drone warfare, Colbert places America's growing dependency on prescription drugs into stark relief. Referencing the all too familiar Cialis ads featuring a couple sitting outdoors in his and hers bathtubs, Colbert reminds us that our "first world problems" pale in comparison to those confronting whole populations living under drones. In entertaining and engaging fashion, Colbert leverages consumer and popular culture for ironic commentary on the American public's complacency toward the targeted killing program.

Well, I'm Not on the List

Critics of satirical news argue these programs promote apathy and cynicism that further compromises an already fragile relationship between engaged citizenship and participatory democracy (Baumgartner & Morris, 2006; Hart & Hartelius, 2007). Proponents of the genre take a different perspective, suggesting instead that parodic news invites viewers to critically interrogate news and official discourse. In this view, programs like *The Colbert Report* represent a form of alternative journalism or oppositional news "one that uses humor to provide the kind of critical challenge that is all but absent in the so-called real news" (Baym, p. 127). While I am sympathetic to this perspective, these assessments routinely overlook high-caliber investigative journalism produced by the independent and alternative news organizations referenced throughout this book. But as we have seen, critical reporting of this sort operates at the margins of conventional journalism and rarely enters into mainstream political discourse. In this final example of satirical perspectives on drones, I want to underscore one of the more important, if least discussed contributions parodic news makes to contemporary public affairs reporting: the amplification of critical perspectives produced through the alternative press and independent reporting.

Promoting his recently released book and documentary film of the same name, *Dirty Wars: The World is a Battlefield*, investigative journalist Jeremy Scahill appeared on *The Colbert Report* to discuss the US drone program.[7] As Geoffrey Baym (2005, 2009) argues, satirical news programs reinvent the traditional late-night talk show interview format in productive and incisive ways. That is to say, in contrast to conventional TV interviews, which invariably generate more heat than light, Colbert's one-on-one discussions produce genuine insight, typically punctuated with good humor. Colbert's discursive strategy in these interviews is to play the foil: "the antagonist who struggles against the interviewee and forces the exchange into a form of dialectical tension" (Baym, 2009, p. 132). While Colbert routinely has the upper hand in these interviews, on this occasion, Scahill makes Colbert think twice about his unwavering support for the targeted killing program.

Acknowledging the danger journalists face when speaking truth to power, Colbert introduces Scahill saying, "It's an honor to be the last person to see him alive." But when Scahill mentions the civilian death toll in "night raids, drone strikes, cruise missile attacks," Colbert feigns ignorance. "What do you mean we're killing a bunch of innocent people? I haven't heard much about that." Colbert's joke underscores US corporate media's role in obscuring the civilian toll of the "surgical" strikes conducted with America's fleet of unmanned aircraft. Undeterred by Colbert's joke, Scahill describes his reporting on "one of the first drone strikes President Obama authorized in Yemen"—an operation that killed 46 people. "Fourteen of them are women, twelve were children. Who was the target of that action? The White House has not provided any information." Scahill's questions are not rhetorical. Rather, they are fundamental inquiries into the legal character and moral legitimacy of the US drone campaign—questions that go largely unasked, and unanswered, in corporate and commercial media.

In a remarkably robust exchange, the sort of deliberative discussion all too rare on American television, Colbert and Scahill debate intelligence gathering, international law, and finally, the extrajudicial assassination of the "radical cleric" Anwar al-Awlaki. Scahill asks, "It doesn't trouble you that the President of the United States is asserting the right to kill American citizens without presenting any actual evidence? Or indicting them at all?" With feigned exasperation, Colbert responds, "Well, I'm not on the list." This lightning-fast comeback gets a laugh from the audience, but as the subject turns to the President's kill list, the joke is on Colbert.

SCAHILL: How do you know?
COLBERT: How do you get on the list?
SCAHILL: I watch this show; you could be on the list.
COLBERT: (Sheepishly) Is it hard to get on the list?
SCAHILL: (Laughing) That's part of the thing, they meet every Tuesday in these Terror Tuesday meetings. …
COLBERT: So you get on the list. Can you get off the list?
SCAHILL: Only from a drone strike it seems.

Again, the audience gasps as a dumbstruck Colbert pauses to reflect on Scahill's response before abruptly ending the interview. "I've got to catch a cab," he explains. In a rare moment on *The Colbert Report*, Scahill shatters Colbert's worldview with black humor that hits a little too close to home for the parodic news host.

Contrasting traditional journalism with parodic news, Geoffrey Baym argues, "satire is the discourse of inquiry, a rhetoric of challenge that seeks through the asking of unanswered questions to clarify the underlying morality of a situation" (2005, p. 267). As we have seen, Western media in general, and US news media in particular, rarely question, let alone challenge, official statements on the precision, accuracy, and efficiency of America's drone program. What's more, voices critical of drone warfare—the perspectives of victims and survivors, drone whistleblowers and anti-war activists—are rarer still. Significantly, these voices offer moral clarity sorely lacking in news and official discourse surrounding the expansive and expanding use of weaponized drones. Humor may seem an unlikely discursive strategy to address these deadly serious issues. And yet, as the preceding discussion demonstrates, political cartoons and parodic news—hybrid genres that integrate news and entertainment discourse—invite readers and viewers to interrogate the moral vacuum surrounding the rise of robotic warfare in the twenty-first century. Of course, satire is no panacea for addressing the vexing issues raised by the proliferation of these killing machines. Nevertheless, satire and irony offer productive avenues to pursue these seemingly intractable questions with candor, insight, and a measure of righteous indignation.

Notes

1. See "Obama Jokes about Killing Jonas Brothers with Predator Drones," *YouTube*, https://www.youtube.com/watch?v=WWKG6ZmgAX4&feature=youtu.be.

2. See "Transcript: Obama's speech against the Iraq War," *NPR*, http://www.npr.org/templates/story/story.php?storyId=99591469.
3. The neoconservative think tank, the Project for a New American Century (PNAC), famously articulated these imperial ambitions in 1997. PNAC members included ten future members of the Bush administration, including Cheney, Rumsfeld, and Wolfowitz.
4. After his 2012 reelection, press reports detailed the Obama administration's plans to develop a so-called "drone rule book" setting out "standards and procedures" for the targeted assassination program in the eventuality that Mitt Romney won the presidency.
5. See "The Word," *The Colbert Report*, http://www.cc.com/video-clips/s3t2y6/the-colbert-report-the-word---two-birds-with-one-drone.
6. See "US drone controversy," *The Colbert Report*, http://www.cc.com/video-clips/1p3who/the-colbert-report-u-s--drone-controversy.
7. See "Jeremy Scahill," *The Colbert Report*, http://www.cc.com/video-clips/vwtsg0/the-colbert-report-jeremy-scahill.

References

Baumgartner, J., & Morris J. S. (2006). *The Daily Show* effect: Candidate evaluations, efficacy, and American youth. *American Politics Research*, *34*, 341–367.

Baym, G. (2005). *The Daily Show*: Discursive integration and the reinvention of political journalism. *Political Communication*, *22*, 259–276.

Baym, G. (2009). Stephen Colbert's parody of the postmodern. In J. Gray, J. P. Jones, & E. Thompson (Eds.), *Satire TV: Politics and comedy in the post-network era* (pp. 124–144). New York, NY: New York University Press.

Conners, J. L. (2005). Visual representations of the 2004 presidential campaign: Political cartoons and popular culture references. *American Behavioral Scientist*, *49*(3), 479–487.

Conners, J. L. (2007). Popular culture in political cartoons: Analyzing cartoonist approaches. *Political Science and Politics*, *40*(2), 261–265.

Day, A. (2011). *Satire & dissent: Interventions in contemporary political debate*. Bloomington, IN: Indiana University Press.

Edwards, D. (2010, May 4). Olberman: Obama 'predator drone' joke was out of bounds. *RawStory*. Retrieved from http://www.rawstory.com/2010/05/olbermann-obama-predator-drone-joke/

Gray, J., Jones, J. P., & Thompson, E. (Eds.). (2009). *Satire TV: Politics & comedy in the post-network era*. New York, NY: New York University Press.

Greenberg, J. (2002). Framing and temporality in political cartoons: A critical analysis of visual news discourse. *Canadian Review of Sociology and Anthropology*, *39*(2), 181–198.

Hart, R. P., & Hartelius, E. J. (2007). The political sins of Jon Stewart. *Critical Studies in Media Communication*, *24*(3), 263–272.

Jenkins, H. (2009). Why Mitt Romney won't debate a snowman. In J. Gray, J. P. Jones, & E. Thompson (Eds.), *Satire TV: Politics and comedy in the post-network era* (pp. 187–209). New York, NY: New York University Press.

Landy, B. (2016, April 30). Larry Wilmore ribs the media at 2016 White House Correspondents Dinner. *Vanity Fair*. Retrieved from http://www.vanityfair.com/news/2016/04/larry-wilmores-best-white-house-correspondents-dinner-burns

Miller, G. (2012, April 30). Brennan speech is first Obama acknowledgement of use of armed drones. *Washington Post*. Retrieved from https://www.washingtonpost.com/world/national-security/brennan-speech-is-first-obama-acknowledgement-of-use-of-armed-drones/2012/04/30/gIQAq7B4rT_story.html

Morris, R. (1993). Visual rhetoric in political cartoons: A structuralist approach. *Metaphor and Symbolic Activity*, 8(3), 195–210.

Schwarz, J. (2016, May 1). DC elite hated Larry Wilmore's drone joke last night, but loved Obama's in 2010. *The Intercept*. Retrieved from https://theintercept.com/2016/05/01/d-c-elite-hated-larry-wilmores-drone-joke-last-night-but-loved-obamas-in-2010/

Solomon, N. (2005). *War made easy: How presidents and pundits keep spinning us to death*. Hoboken, NJ: John Wiley & Sons.

Teather, D. (2004, March 26). Bush jokes about search for WMD, but it's no laughing matter for critics. *The Guardian*. Retrieved from https://www.theguardian.com/world/2004/mar/26/usa.iraq

Thompson, E. (2009). Good demo, bad taste: *South Park* as carnivalesque satire. In J. Gray, J. P. Jones, & E. Thompson (Eds.), *Satire TV: Politics and comedy in the post-network era* (pp. 213–232). New York, NY: New York University Press.

Wall, T. (2016). Ordinary emergency: Drones, police, and geographies of legal terror. *Antipode*, 48(4), 1122–1139.

CONCLUSION

Twenty-First Century Empire and Communication

Our enemies are innovative and resourceful, and so are we.
They never stop thinking about new ways to harm our country
and our people, and neither do we.

—George W. Bush[1]

In this final chapter, I examine the rise of drone warfare in light of media theorist Harold Innis's (1995, 2007) seminal work on empire and communication. Specifically, this discussion applies and extends Innis's insights on the bias of communication to a critical analysis of the targeted killing program initiated under President George W. Bush and expanded by his successor, Barack Obama. Throughout, I contend weaponized drones embody a long-standing paradox of American culture: the impulse to collapse the geographical distance between the United States and other parts of the globe, while simultaneously magnifying the cultural differences between Americans and other peoples and societies. Doing so, the chapter demonstrates the relevance of Innisian approaches to the relationship between technology and culture articulated through this new kind of war.

The chapter is organized into three sections. The first offers a brief description of the aerial, navigational, imaging, and computational technologies employed by America's fleet of weaponized drones. Throughout, I highlight the networked intelligence that distinguishes armed drones from other weapons platforms. This section also describes the operational logistics required to launch these aircraft, surveil their targets, and fire their missiles. Contrary to media portrayals and popular belief, drone missions require more than a handful of operators. Rather, combat patrols routinely involve three or four unmanned aircraft, requiring upwards of 150 personnel, including ground crews, remote pilots, sensor operators, intelligence analysts, and legal advisors (Chatterjee, 2015). These flight teams constitute the operational component of what the US military aptly describes as the "kill chain" (Cockburn, 2015). Innis's insights into the relationship between communication technologies and the rise of administrative classes—what he describes as monopolies of knowledge and power—can usefully be applied to "the assassination complex" (Scahill, 2015) that authorizes and conducts drone strikes.

The second section employs Innis's space-time dialectic to examine the rise of drone warfare. According to Innis, communication technologies manifest and express certain tendencies—what he called biases—that enable or constrain the ability of organizations, institutions, and empires to control space (territory) and time (continuity). Innis's historical analysis reveals how speech and the oral tradition emphasize the temporal duration of social organization, thereby facilitating the establishment of long-lasting and sustainable communities. Conversely, the written word, followed by print and later electronic communication, emphasize the spatial extension of social organization, thereby facilitating territorial growth and administrative control over enormous expanses. This expansionist tendency deeply troubled Innis. According to James Carey and John Quirk (1992), Innis "placed the 'tragedy of modern culture' in America and Europe upon the intrinsic tendencies of both printing press and electronic media to reduce space and time to the service of a calculus of commercialism and expansionism" (pp. 133–134).

The third section considers the moral and political implications of drone warfare in light of Innis's thinking about technology and culture. As Catherine Frost (2003) reminds us, "There is a powerful moral voice speaking in his communications work, and it carries a message about the achievements possible in a good society and the brutalizing effects in a bad one." Writing in the midst of a Cold War that threatened to annihilate humankind, Innis's work took on great urgency. Were he alive today, Innis would no doubt condemn the depravity of killing by remote control. Here, then, we can appreciate the value of taking a discursive turn to Innis's approach to empire and communication (Angus, 1998). That is to say, Innis understood that the bias of communication could only be controlled through deliberative politics.

> The genius of social policy, he thought, was to serve the demands of both time and space; to use one to prevent the excesses of the other: to use historicism to check the dreams of reason and to use reason to control the passions of memory. (Carey, pp. 160–161)

In short, Innis fully appreciated the role of language and culture in shaping technology—and viewed his scholarship as an intervention into the politics of technological development.

The chapter concludes with some thoughts on Innis's legacy and its relevance for both communication and American studies in the age of drone warfare. As Megan Mullen notes, "One might well claim that the American empire has been Innis's best theorized, even though he did not live to witness its most revealing manifestations" (2009, p. 177). Simply put, the US drone program is but the latest manifestation of American empire. In an effort to mitigate the self-destructive values and perspectives fostered by space-biased cultures, then, we must be wary, as Innis was, of inscribing new technologies with the sublime (Carey & Quirk). Doing so, especially in the American context, has wrought a great many engineering and electrical wonders: from the railroad and the Brooklyn Bridge to the electric light and the moon landing. More ominously, the American fascination with new forms of what David Nye (1994) calls "the dynamic sublime" led to the development of atomic weapons that so troubled Innis at middle of the last century—and which likewise ushered in the era of drone warfare at the dawn of new millennium.

It's About the Data Link, Stupid

In news and popular culture, the growing fascination with drones rests in large measure on the futuristic design of these aircraft and the lethal payload they carry. As we have seen, for entrepreneurs and commercial drone manufactures this public perception is both a blessing and a curse. But for students of communication, the drone's true innovation lies elsewhere. Speaking to *The Atlantic* magazine, retired Air Force major general James Poss, who supervised the transformation of the Predator from a surveillance vehicle to a weapons platform, put it plainly: "It's about the data link, stupid" (Bowden, 2013, p. 62). Absent the array of information gathering and processing technologies used to conduct US drone strikes, these unmanned planes are something of a throwback to the propeller driven aircraft of the early twentieth century. Thus, these high tech killing machines integrate modern transportation and communication technologies: essential ingredients for the command and control requirements of empire, both ancient and modern. As Wayne Cocroft observes, "The Roman Empire developed a sophisticated system of roads, imperial messengers and, along its frontiers, a network of signal stations" (2013, p. 65). Likewise, the US drone fleet—the tip of America's imperial spear—relies upon a network infrastructure that connects these remotely piloted aircraft with far-flung launch and recovery units, signal processing centers, and ground control stations (GCS) via radio, fiber optics, and telecommunication satellites.

Made of lightweight composite materials that enable the aircraft to loiter above a target for hours at a time, the distinguishing physical characteristic of the Predator drone—and its larger more heavily armed counterpart, the Reaper[2]—is the vehicle's bulbous nose. Housing an array of sensors, actuators, antennae, and cameras, this section of the aircraft not only allows pilots to fly these aircraft from halfway around the world, but also gives these vehicles an unprecedented ability to combine long duration intelligence, surveillance, and reconnaissance (ISR) missions with the capacity to take lethal action. Specifically, these operations include launch and recovery units using line-of-sight C-band radio signals operating in or nearby the aircraft's theater of operation in the skies above Afghanistan, Iraq, Pakistan, Yemen, and Somalia. Once airborne, a ground control station in the United States assumes control of the vehicle using Ku-band satellite signals. Data and images collected by these unmanned planes are processed at the US Air Force base in Ramstein, Germany and subsequently relayed across the Atlantic Ocean to the United States via fiber optic cables.

CIA operations originate from the intelligence agency's headquarters in Langley, Virginia, while Creech Air Force Base in Nevada is home to the US Air Force's best-known ground control station.[3] Meanwhile, Special Operations Forces (SOF) and other ground troops have real-time access to information and imagery collected by drones flying overhead. In turn, authorized military personnel, intelligence analysts, and civilian leaders around the world access this data via the Distributed Common Ground System (DCGS): what journalist Andrew Cockburn describes as a "little-known global network" that acts as "the central nervous system funneling, collating, and sharing unimaginable quantities of imagery and electronic information" (2015, p. 3).

Coupling intelligence gathering, data analysis, target selection, and rapid strike capabilities in this fashion, drones are the embodiment of what Pentagon strategists describe as "network centric warfare" (Kometer, 2003, p. 36).

From an Innisian perspective, drones exploit the twin revolutions in satellite navigation and digital communication to annihilate space and time in the service of military expansionism. Tracing imperial history through the ages, Innis came to view media and communication technologies as "the staples of empire" (Watson, 2007, p. 14). Improvements in navigation and communication cultivated an expansionist mindset increasingly reliant on the projection of force over vast territories. Describing the rise and fall of ancient dynasties, Innis found "military organization essential to the expulsion of the invaders became the basis of expansion and growth of an Egyptian empire. Protectorates were established beyond the borders as a means of economy in the use of soldiers and in administration costs" (Watson, p. 40). Similarly, in the aftermath of the Second World War, the United States emerged as a global superpower whose reach and influence extended across Europe and the Pacific Rim as never before. Since that time, America's imperial ambitions have been inextricably linked to the rise of what President Dwight D. Eisenhower (1961) famously described as the "military-industrial complex": a self-perpetuating alliance between the military, industry, academia, and the political class, that not only assured US global hegemony but also secured the political-economic dominance of arms manufactures.

In the aftermath of the attacks of September 11, 2001, the United States embarked on a new round of imperial expansion, including the repurposing of existing military bases and the construction of new ones, to facilitate what war planners call full spectrum dominance: the ability to control every dimension of the battlefield: land, air, and sea; as well the virtual terrain of information space. As a result, the so-called defense industries—aeronautics firms, electronics and telecommunication companies, weapons manufactures, private contractors and related corporate interests—reaped enormous profits over the course of America's decades-long war on terror (Turse & Engelhardt, 2012). Notwithstanding the chronic failure of US news organizations to acknowledge, much less discuss American empire, an emerging constellation of drone bases has not gone completely unnoticed. For instance, in 2011 the *Wall Street Journal* reported: "The military has reopened a base for the unmanned aircraft on the island nation of Seychelles to intensify attacks on al Qaeda affiliates, particularly in Somalia" (Barnes, 2011). More recently, investigative journalist Nick Turse (2016) identified no fewer than 60 such bases across Africa—including a single $100 million drone base under construction in Niger. Like earlier empires that built transportation and communication infrastructure for purposes of administration and control over vast territories, these postcolonial outposts enable the United States to project force across the globe: albeit with little public knowledge at home; but not without the support of local, frequently despotic regimes. Writing for the independent news site *TomDispatch*, Turse (2015) notes:

> These bases, camps, compounds, port facilities, fuel bunkers and other sites can be found in at least 34 countries—more than 60% of the nations on the continent—many of them corrupt, repressive states with poor human rights records.

Thus, the "sorrows of empire," to borrow Chalmers Johnson's (2004) haunting characterization of US militarism, encompass far more than the eclipse of American democracy and the nation's economic subservience to merchants of death. Like Innis before him, Johnson recognized and cautioned against the brutalizing effects of space-bound societies evident in twenty-first century American empire and its unrelenting assault on tribal Islam.

This ring of drone bases increasingly relies on the National Security Administration's (NSA) global surveillance infrastructure to "find, fix, and finish" (F3 in military shorthand) targets of US drone strikes. To that end, the NSA employs two tactics: the first uses the intelligence agency's phone-tracking capabilities to geolocate the cell phone SIM cards of potential targets. NSA furnishes this information to either the CIA or the Joint Special Operations Command's (JSOC) High Value Targeting task force for subsequent "kinetic military action."[4] Significantly, this technique does not monitor or collect the content of cell phone conversations; rather, geolocation relies on the metadata associated with cell phone activity. The second tactic equips drones and other aircraft with so-called virtual base tower receivers—code name Gilgamesh after the Ancient Sumerian demigod—that mimic cell phone towers, effectively turning a suspect's phone into a tracking device. As we shall see, the drone program's reliance on signals intelligence (SIGINT) raises profound moral questions. For present purposes, the sheer volume of metadata gleaned from cell phones—as well as computers, wireless routers and other electronic equipment—is noteworthy.

In the words of a former drone operator, this data requires "millions of dollars and millions of man hours" to assess and interpret (Scahill & Greenwald, 2014). In turn, this process yields material collected in target information folders on suspected terrorists; ultimately this information is distilled into the aforementioned baseball cards used by the President and his national security team to select individuals for targeted assassination. Once SIGINT finds a target, the "unblinking stare" (Coll, 2014) of overhead drones employ imaging technologies, including daylight and infrared cameras, to fix a target in a specific location until such time as JSOC, the CIA or other high-ranking officials authorize a drone strike to finish the target. In short, intelligence analysts are but one link in a kill chain that extends from ground control stations to the Oval Office.

What investigative journalist Jeremy Scahill (2015) describes as "the assassination complex" resembles the bureaucracies Innis observed coalescing around each new medium of communication. According to Innis, these administrative classes constitute monopolies of knowledge and power that exert enormous influence over social, economic, and political life. As we noted in Chapter 9, the distributed intimacy engendered by drone warfare is asymmetrical. US personnel come to know intimate details about their quarry, while potential targets know little more than the fact that their lives are under constant surveillance (Gregory, 2011). Conversely, the clandestine nature of the drone program, including specifics regarding how targets are selected, represents yet another dimension of this monopoly of knowledge. Outside of the President's national security team, few members of Congress, let alone the American people, have even a rudimentary understanding of the inner-workings of the targeted killing program. As Innis notes of Egyptian empire, "Complexity favoured increasing control under a monopoly of priests and the confinement of knowledge to special classes" (2007, p. 44). So too, the secrecy and sociotechnical complexity of the drone program constitutes a special class of "hi-tech assassins" (Cockburn). Put differently, in twenty-first century American empire, the President's self-granted authority to act as judge, jury, and executioner—coupled with what a former drone operator described as the "godlike power" of killing by remote control[5]—bears a striking resemblance to the religious and quasi-religious consolidation of power Innis identifies with monopolies of knowledge.

Drones and the Problems of Time and Space

In his final years—at the middle of the last century—Harold Innis lamented the shortsighted, and to his mind, self-defeating tendencies of space-biased cultures, and cautioned against the diminishing importance of time, reflection, and historical perspective. All of which begs the question: What would Innis make of twenty-first century empire and communication? For Gatean Tremblay the answer is clear:

> Computers and telecommunications networks have strengthened this trend significantly, to the point that one wonders if we are not witnessing the emergence of a new form of imperialism, which would not necessitate, in most cases, the establishment of a sprawling political administration. (2012, p. 573)

We can discern this new form of imperialism in the space and time annihilating capabilities of weaponized drones described above. Leveraging leading-edge aeronautics with advanced optics, data processing, and networked communication, drones represent an archetypal "space-biased" technology. Furthermore, the ability of remote operators and others to monitor, select, and strike targets from half a world away, and in real time,[6] epitomizes the "pernicious neglect of time" Innis analyzed in his later writing (Compton & Comor, 2007, p. 47). This growing contempt for time facilitates and promotes a state of perpetual war.

Indeed, despite the much-heralded success of the US drone campaign in news and official discourse, the so-called war on terror proceeds with no end in sight. Nearly two decades on, American forces continue to conduct military operations across the Greater Middle East, with little to show for it other than increasing tensions and volatility. The possibility that this regional instability is due, in large measure, to American militarism rarely enters mainstream political discourse. Nonetheless, the targeted killing program has failed to defeat jihadi movements and may in fact serve as a recruitment tool for extremist groups across the region; on a par with the radicalization of Muslims in the wake of revelations of indefinite detention and torture of suspected terrorists at the US detention facility at Guantanamo Bay, Iraq's infamous Abu Ghraib prison, and dozens of CIA "black sites" around the world.

Rather than eliminate potential threats, US drone strikes fuel popular resentments and swell the ranks of local insurgencies. This is a recipe for endless war: a corrosive and counterproductive strategy that represents a genuine existential threat to American civilization. Writing in the early 1950s, Innis argued: "Losing touch with the problems of time, [the state] has been willing to engage in wars to carry out immediate objectives" (1995, p. 76). From an Innisian perspective, then, the short-term tactical victories afforded by drone strikes that kill terrorist leaders overshadow the long-term strategic implications of waging perpetual war.

Killing by remote control has become so commonplace that analysts sometimes characterize the program with a callous but revealing euphemism: mowing the lawn.[7] The domesticating logic behind this metaphor is plain to see: an otherwise mundane household ritual serves to normalize and routinize drone assassination. This seemingly innocuous figure of speech belies the depraved indifference for human life cultivated by drone warfare and underscores a growing indifference for time Innis associates with space-biased cultures. Simply put, the euphemism "mowing the lawn" performs the important ideological work of promoting, while simultaneously trivializing, a state of perpetual war.

In addition to normalizing endless war, unmanned aircraft and the associated surveillance and computational technologies used to conduct robotic warfare work in effect to weaponize time. Doing so, this new kind of war exacerbates historical processes, which, Innis argued, "destroyed a sense of time" (1995, p. 86). Consider the US Air Force's Gorgon Stare IRS system. Grouping the imagery taken by as many as ten remote planes flying simultaneously over a target area, the Gorgon Stare enables analysts to continuously monitor and record activity throughout an entire city. "Computers store these moving images so that analysts can *dial back to a particular time* and place and zero in, or mark certain individuals and vehicles and instruct the machines to track them over time" (emphasis added, Bowden, p. 63). This preoccupation with prior knowledge—something of an obsession in the post 9/11 American psyche—is antithetical to the importance of time to the stability, continuity, and duration of societies Innis observed in his historical analysis of media and communication.

Likewise, the use of so-called signature strikes, which target unidentified groups of "military-aged men" based solely on their geographic location, physical attributes, and patterns of behavior, reveals an utter disregard for time emblematic of space-biased technologies and societies. In this respect, signature strikes dramatically reveal "an undue obsession with the immediate" (Innis, 1995, p. 61) that all too often claims the lives of innocent civilians. Unlike targeted assassination, this tactic does not require presidential approval. Rather military personnel monitor, surveil, and frequently assault targets whose actions bear the "signature" of militant activity, without any evidence of violent or criminal activity. Small wonder, then, that critics liken signature strikes to "pre-crime"—a preemptive law enforcement technique that incarcerates people suspected of committing some future crime (Calhoun, 2016).

As depicted in Steven Spielberg's 2002 thriller *Minority Report*, based on a short story of the same name by acclaimed science-fiction writer Philip K. Dick, would-be criminals are apprehended *before* they break the law. But in contemporary law enforcement and military operations, the philosophy and practice of pre-crime has transmogrified from the stuff of dystopian fiction to the harsh reality of lived experience (Smith, 2016). From an Innisian perspective, pre-crime reduces time to a commodity, subject to manipulation and control by monopolies of knowledge and power. Unlike pre-crime, however, signature strikes forego capture altogether and instead kill suspects without charge or trial. Upending the temporal order of criminal investigations, pre-crime attempts to predict and control criminal activity before it occurs. So too, signature strikes violate any notion of due process—a time-honored legal tradition—and instead targets groups of people for execution based on actions and behaviors observed from "eyes in the sky."

The US government classifies anyone killed in a drone strike as "enemies killed in action" (EKIA) without knowing their identities. Rarely does the US acknowledge innocent victims and only in posthumous fashion. "Shoot first and ask questions later"—a quintessentially American sentiment—seems an apt description for this tactic. The cruel irony in all of this reveals an utter contempt for time: a barbarous and self-defeating present-mindedness that Innis would doubtless find appalling, but not at all surprising, given American empire's reliance upon, and fascination with, the sublimity of space and time annihilating technologies.

Significantly, the bias of space-oriented media and communication is evident in the official language used to justify drone assassination. In Chapter 3, we noted that President Obama's Office of Legal Counsel developed a "broader concept of imminence" to justify killing individuals, without evidence, who might, at some unspecified point in time, be involved in an attack on US interests and citizens. Critics decried this abuse of language in legal arguments asserting the President's authority to execute suspected terrorists. The elasticity of this new definition, critics contend, not only defies common understanding and usage of the word. It sets a troubling legal precedent, one that epitomizes a profound neglect for time. Without putting too fine a point on it, this lexical sleight of hand may have significant and unforeseen ramifications in the years to come as other nations and non-state actors deploy weaponized drones of their own (Kreps & Zenko, 2014). Such is the trap of present-mindedness Innis cautioned against. In sum, the lexical strategy of redefining the word imminent to provide legal justification for targeted killing does more than rob the word of its meaning. Rather it is the discursive embodiment of the values, outlooks, and aspirations expressed by and constitutive of space-biased cultures. All of which underscores the crucial and decisive role discourse plays in shaping, as well as reflecting, technological development.

Innis and the Discursive Turn

With his focus on the material features of media and communication technology—its relative lightness or heaviness, its ephemeral or durable character—Innis's historical analysis is sometimes criticized as reductionist and determinist. These critiques overlook or fail to appreciate Innis's attention to the varied forces and conditions—economic and cultural, geographic and political—that shape and inform technological innovation. Foregrounding the significance of space and time to the growth and development of human societies, Innis viewed the relationship between communication and social organization as dialectical and contingent, and made this the centerpiece of his study. As Ian Angus (1998) puts it, "The concepts of time and space are used to describe the constitutive power of media of communication in constructing and maintaining society."

Furthermore, Innis's historical analysis suggests that communication technologies are both a reflection and an expression of a given society's bias toward its maintenance in time, or its expansion in space. In other words, while technologies have inherent characteristics, they are nonetheless malleable and, Innis believed, ought to be developed in a relational fashion, for purposes of serving society's interest across both time and space. Viewed in this light, Innis's work is at once descriptive and prescriptive. On the one hand, Innis identifies and celebrates past societies that achieved such a balance; on the other hand, he pleads for modern societies to strike a similar balance between these two cultural biases (Innis, 1995).

Writing in *Empire and Communications*, Innis (2007) observes that human cultures "flourish under conditions in which civilization reflect the influence of more than one medium and in which the bias of one medium toward decentralization is offset by the bias of another medium toward centralization" (p. 27). By contrast, an over emphasis on either dimension leads to monopolies of knowledge and power that undermine social stability and security. Equally important, as Catherine Frost (2003) observes, Innis recognized the moral implications of his analysis: "More was at stake than the staying power of civilizations. The civility of civilizations, their worthiness as articulations of the human community mattered as well." Emblematic of the most repellent tendencies and aspirations of space-biased cultures, the use of drones to kill people from half a world away raises profound moral questions precisely because they epitomize the incivility, the cowardice—indeed the barbarism—of this new kind of war.

Of course, saying this contradicts news and official discourse that portrays these killing machines as sophisticated weapons capable of hitting their targets with pinpoint accuracy and surgical precision. But as we've noted throughout, drone survivors, investigative journalists, human rights lawyers, and anti-war activists produce and circulate alternative and oppositional discourses that challenge this dominant order of discourse. Consider the words of a former drone operator regarding the aforementioned geolocation targeting practices used to conduct and coordinate US drone strikes. Speaking to *The Intercept*, the unnamed source claims that insurgents are well aware of the NSA tracking system and have developed tactics to conceal or disguise their identities. "They would do things like go to meetings, take all their SIM cards out, put them in a bag, mix them up, and everybody gets a different SIM card when they leave. That's how they confuse us." (Scahill & Greenwald).

The moral hazard of what this whistleblower describes as "death by metadata," rarely acknowledged in such candid terms, undermines repeated claims by the Obama administration that the US conducts drone strikes only when individuals have been identified and that there is "near certainty" that innocent civilians will not be killed or injured. But as the former drone operator concludes, more likely than not, the person with the phone at the time of an assault is not the intended target.

> People get hung up that there's a targeted list of people. It's really like we're targeting a cell phone. We're not going after people—we're gong after their phones, in the hopes that the person on the other end of that missile is the bad guy. (Scahill & Greenwald)

This same report underscores the moral discrepancies between US intelligence gathering practices before and after a drone strike. Prior to an assault, the kill chain relies almost exclusively on SIGINT: the sort of intelligence gathering techniques insurgents have learned to compromise and confound. Despite the moral if not the tactical advantage of having eyes and ears on the ground, as well as in the sky, human intelligence (HUMINT) rarely contributes to the target selection process. In contrast, Caitlin Hayden, a spokesperson for the National Security Council (NSC), described after action assessments this way:

> After any use of targeted lethal force, where there are indications that civilian deaths may have occurred, intelligence analysts draw on a large body of information—including human intelligence, signals intelligence, media reports, and surveillance footage—to help us make informed determinations about whether civilians were in fact killed or not. (Scahill & Greenwald)

The twisted logic of confirming the presence of civilian casualties *after* an assault has taken place raises profound moral questions: precisely the sort of humanist concerns that motivated Innis's communications work at the dawn of the Cold War (Frost).

To that end, and in keeping with his training as an economist, Innis sought to strike a balance between time and space biased media. The moral and political shortcomings of industrial societies—especially the problems of "mechanized knowledge" Innis found most troubling—find a counterbalance in the dialogic character of the oral tradition.

> Innis claims that orality was a stabilizing influence in civilization in the past, though this has not been adequately enough understood. In the present we need to recover and extend orality in order to develop greater stability in time and this is the healing intention of Innis's theory—to restore balance where balance has been disturbed. (Angus)

Contrary to criticism that his work suffers from rigid technological determinism, Innis understood full well that the balance he sought was not a matter of technological innovation, but rather of politics and culture (Acland, 2006). Nowhere is this more evident than in the discursive struggle over America's use of armed drones in its decades-long war on terror. As documented throughout this book, political and military leaders—working in tandem with their enablers in the corporate news media—attempt to dominate policy discussions through secrecy, selective leaks, and sanitized press coverage of US drone strikes. Conversely, elements of civil society exercise power and influence policy through all manner of discursive practices—investigative reporting, whistleblowing, street protests, and graphic arts, to name a few.

None of which is to ignore or underestimate the power differentials operating across these orders of discourse. Still, this discursive struggle has tangible effects. For instance, in July 2016, following years of litigation and agitation on behalf of innocent drone victims and their families, President Obama issued an executive order calling for greater accountability for civilian casualties of US drone strikes. Three months later, a coalition of human rights groups followed up on the new policy and sent a letter to Barack Obama identifying ten drone strikes "requiring investigation and acknowledgement" of civilian casualties (Fulton, 2016). These efforts to introduce a measure of transparency in America's drone program underscore the significance of creating a discursive space to debate the wisdom, efficacy, and morality of drone warfare. Equally important, the discursive struggle surrounding the US drone program offers a stark political lesson on the relationship between technology and culture. As James Carey puts it: "The bias of technology can be controlled only by politics, by curtailing the expansionist tendencies of technological societies and by creating avenues of democratic discussion and participation beyond the control of modern technology" (Carey & Quirk, p. 136).

Drones: The New American Sublime

It is tempting to view the so-called war on terror in terms of a "clash of civilizations" between the Western and Muslim worlds (Lewis, 2002). The appeal of this narrative hinges on longstanding and seemingly intractable tensions over religion and politics. But as Akbar Ahmed observes, a significant and frequently overlooked aspect of this analysis lies in the recognition that "the clash of civilizations is not the inevitable outcome of history, but a sign that communication and exchange has broken down" (2013, p. 304). Ahmed's insight echoes a similar sentiment Harold Innis expressed some 65 years ago: "Enormous improvements in communication have made understanding more difficult" (1995, p. 31).

This paradox is all the more startling in light of the multiplicity of communication forms and technologies employed by state and non-state actors alike in their respective struggles to win hearts and minds or, conversely, to strike terror in the hearts of their adversaries. Militant groups like al-Qaeda and, more infamously, ISIS, issue threats and document all manner of atrocity through social media and related platforms. All to the consternation and indignation of Western viewers (Rowley, 2014). Conversely, the Department of Defense posts footage of US drone strikes to YouTube for purposes of "promoting UAVs domestically and 'enlightening' our enemies" (Thompson, n.d.). While the success of this "drone porn" in dissuading would-be extremists from taking up arms remains unclear, these clips generate considerable interest online (Stahl, 2013).

As I have sought to demonstrate throughout this book, the rhetoric of the technological sublime pervades much of our language and a great deal of our thinking about drones. The ability to wage war from a world away; to strike at adversaries while keeping US forces relatively free from harm has made war easy: a sublime state of perpetual conflict. In the American imagination, the technological sublime represents progress, ingenuity, and unrivaled mastery over the natural world. Curiously, the electric sublime—replete with promises of liberty, fraternity, and prosperity—finds one of its most vocal proponents in Marshall McLuhan, who was deeply influenced by Innis's work. But Innis was far more skeptical of the rhetoric of the electric sublime.

As Carey and Quirk point out, Innis "disputed the notion that electricity would replace centralization in economics and politics with decentralization, democracy, and a cultural revival" (p. 133). Drawing upon Innis's clear-eyed assessment of more recent developments in computational, navigation, and communication technologies, Carey continues: "What we are witnessing is another increase in the scale of social organization based upon electronic communication" (p. 170). Put in more theoretical terms, integrating elements of both the dynamic and electric sublime, drones represent a rather ominous development in an ongoing historical process of time-space compression that collapses geographic distance, frequently exacerbating unequal relations of social, economic, and political power (Harvey, 1990).

In conclusion, Innis's work offers an invaluable heuristic tool for understanding the relationship between the technological sublime and the growth and expansion of American empire. The dynamic sublime of the railroad, coupled with the electric sublime of the telegraph, were instrumental in colonizing a continent and its Indigenous peoples. Led by the drone program, this latest round of American imperial expansion likewise threatens tribal Islam—oral cultures, not unlike the native peoples of North America, deeply rooted in place and tradition. All of which is to suggest that the sublime frequently (and purposefully) obscures the brutalizing effects of space-binding societies and technologies.

In the American context, the sublime represents far more than a quasi-religious feeling or sensation. Rather, it is a way of thinking that inscribes technologies and social relations in discernable, replicable, perhaps even predictable ways; what Charles Acland describes as "grooves."

> It behooves us to examine seriously the lasting consequences of the grooves in social existence as scratched out by policy programs. In the end, we need to ask not only about the decisions taken, but *what manner of cultural coherence* [emphasis added] is being fabricated, fortified and advanced by said decisions. (2006, p. 176)

Innis's work poses fundamental questions regarding policy making in the twenty-first century, especially, as I have suggested throughout this book, regarding the US drone program and the attendant arms race in militarized drones across the globe. In keeping with the spirit of Innis's critique, my modest aim here is to enlarge and enhance a public conversation about robotic warfare. Equally important, Innis's moral voice demands that we consider the broader implications of such cultural coherence in a society increasingly enamored with all manner of drones: the new American sublime.

Notes

1. Bush made this unintentionally revealing comment on August 5, 2004 during a signing ceremony for a $417 billion defense spending bill. See "President gaffes in terror speech," *BBC News*, http://news.bbc.co.uk/2/hi/americas/3541706.stm.
2. Billed as the United States' most lethal unmanned aircraft, the Reaper carries laser guided Hellfire missiles, much like the Predator, as well as two 500 lb. "smart bombs." In contrast, the high altitude Global Hawk is an unarmed surveillance platform.
3. As CIA missions increased in degree and frequency throughout the Obama administration, these operations were outsourced to Air Force personnel at Creech.
4. Commonly used by politicians and media spokespeople, this phrase is another in a long line of euphemisms for the use of violent force by US military and paramilitary units.
5. See Scahill (2015).
6. As Andrew Cockburn (2015) observes, there is in fact a 2–5 second delay or "latency" owing to the vast distance data travels from an airborne drone, to orbiting communication satellites, then via fiber optic cable from the US Air Force base in Ramstein, Germany, across the Atlantic Ocean and on to command and control stations across the continental United States.
7. For example, see this opinion piece by former US Congressman Dennis Kucinich, "Obama administration must account to Congress for targeted assassinations," *The Guardian*, https://www.theguardian.com/commentisfree/2012/nov/16/obama-administration-account-congress-targeted-assassinations.

References

Acland, C. R. (2006). Harold Innis, cultural policy, and residual media. *International Journal of Cultural Policy*, *12*(2), 171–185.

Ahmed, A. (2013). *The thistle and the drone: How America's war on terror became a global war on tribal Islam*. Washington, DC: Brookings Institution Press.

Angus, I. (1998). The materiality of expression: Harold Innis's communication theory and the discursive turn in the human sciences. *Canadian Journal of Communication, 23*(1). Retrieved from http://www.cjc-online.ca/index.php/journal/article/%20view/1020/926

Barnes, J. E. (2011, September 21). US expands drone flights to take aim at East Africa. *Wall Street Journal*, p. A1.

Bowden, M. (2013, September). The killing machines: How to think about drones. *The Atlantic*, 58–70.

Calhoun, L. (2016, May 9). What's worse than the department of pre-crime? The US drone program. *Antiwar.com*. Retrieved from https://www.antiwar.com/blog/2016/05/09/whats-worse-} than-the-department-of-pre-crime-the-us-drone-program/

Carey, J. W. (1992). *Communication as culture: Essays on media and society*. New York, NY: Routledge.

Carey, J. W., & Quirk, J. J. (1992). The mythos of the electronic revolution. In J. W. Carey *Communication as culture: Essays on media and society* (pp. 113–141). New York, NY: Routledge.

Chatterjee, P. (2015, July 12). No lone rangers in drone warfare. *TomDispatch*. Retrieved from http://www.tomdispatch.com/blog/176022/tomgram:_pratap_chatterjee,_no_lone_rangers_in_drone_warfare

Cockburn, A. (2015). *Kill chain: The rise of the high-tech assassins*. New York, NY: Henry Holt.

Cocroft, W. D. (2013). The archaeology of military communications. *Industrial Archaeology Review, 35*(1), 65–79.

Coll, S. (2014, November 24). The unblinking stare. *The New Yorker*. Retrieved from http://www.newyorker.com/magazine/2014/11/24/unblinking-stare

Compton, J. R., & Comor, E. (2007). The integrated news spectacle, Live 8, and the annihilation of time. *Canadian Journal of Communication, 32*, 29–53.

Eisenhower, D. D. (1961). Farewell address. *Eisenhower Presidential Library*. Retrieved from https://www.eisenhower.archives.gov/research/online_documents/farewell_address.html

Frost, C. (2003). How Prometheus is bound: Applying the Innis method of communications analysis to the internet. *Canadian Journal of Communication, 28*(1). Retrieved from http://www.cjc-online.ca/index.php/journal/article/view/1338/1396

Fulton, D. (2016, October 7). Human rights groups to Obama: Time to follow through on drone promises. *Common Dreams*. Retrieved from http://www.commondreams.org/news/2016/10/07/human-rights-groups-obama-time-follow-through-drone-promises

Gregory, D. (2011). From a view to a kill: Drones and late modern war. *Theory, Culture & Society, 28*(7–8), 188–215.

Harvey, D. (1990). *The condition of postmodernity*. Oxford: Blackwell.

Innis, H. A. (1995). *The bias of communication*. Toronto, Canada: University of Toronto Press.

Innis, H. A. (2007). *Empire and communications*. Toronto, Canada: Dundurn Press.

Johnson, C. (2004). *Blowback: The costs and consequences of American empire*. New York, NY: Henry Holt.

Kometer, M. (2003). The strategy of control: Centralized vs. decentralized control of US airpower. *Defence Studies, 3*(2), 36–63.

Kreps, S., & Zenko, M. (2014). The next drone wars. *Foreign Affairs*, *93*(2), 68–79.

Lewis, B. (2002). *What went wrong? The clash between Islam and modernity in the Middle East*. New York, NY: Harper Perennial.

Mullen, M. (2009). Space bias/time bias: Harold Innis, empire and communication. *Technology and Culture*, *50*(1), 175–186.

Nye, D. E. (1994). *American technological sublime*. Cambridge: MIT Press.

Rowley, C. (2014, September 30). Beheadings v. drone assassinations. *Consortium News*. Retrieved from https://consortiumnews.com/2014/09/30/beheadings-v-drone-assassinations/

Scahill, J. (2015, October 15). The drone papers: The assassination complex. *The Intercept*. Retrieved from https://theintercept.com/drone-papers/the-assassination-complex/

Scahill, J., & Greenwald, G. (2014, February 10). The NSA's secret role in the US assassination program. *The Intercept*. Retrieved from https://theintercept.com/2014/02/10/the-nsas-secret-role/

Smith IV, J. (9 October 2016). Crime-prediction tool PredPol amplifies racially biased policing, study shows. *Mic*. Retrieved from https://mic.com/articles/156286/crime-prediction-tool-pred-pol-only-amplifies-racially-biased-policing-study-shows#.wyQJ7nmzJ

Stahl, R. (2013). What the drone saw: The cultural optics of the unmanned war. *Australian Journal of International Affairs*, *67*(5), 658–674.

Thompson, K. (n.d.). Drone porn: The newest YouTube hit. *Huffington Post*. Retrieved from http://www.huffingtonpost.com/keith-thomson/drone-porn-the-newest-you_b_407083.html

Tremblay, G. (2012). From Marshall McLuhan to Harold Innis, or from the global village to world empire. *Canadian Journal of Communication*, *37*(4). Retrieved from http://www.cjc-online.ca/index.php/journal/article/view/2662

Turse, N. (2015, November 17). Does eleven plus one equal sixty? AFRICOM's new math, the US base bonanza and 'scarier' times ahead in Africa. *TomDispatch*. Retrieved from http://www.tomdispatch.com/post/176070/tomgram:_nick_turse,_america's_empire_of_african_bases/

Turse, N. (2016, September 29). US military is building a $100 million drone base in Africa. *The Intercept*. Retrieved from https://theintercept.com/2016/09/29/u-s-military-is-building-a-100-million-drone-base-in-africa/

Turse, N., & Engelhardt, T. (2012). *Terminator planet: The first history of drone warfare 2001–2050*. Lexington, KY: Dispatch Books.

Watson, A. J. (2007). Introduction. *Empire and communication*. Toronto, Canada: Dundurn Press.

INDEX

A

B

C

D

E

F

F

G

H

I

J

K

L

M

N

O

P

T

U

V

W

Y

Z